The Tuscany Dialogues

NEW PARADIGM BOOKS OF
THE LASZLO INSTITUTE OF NEW PARADIGM RESEARCH
Kingsley L. Dennis, Series Editor

AN ECOLOGY PRIME™ PUBLICATION

The Tuscany Dialogues

The Earth, Our Future, and the Scope of Human Consciousness

*A Six-Day Conversation Between
Ervin Laszlo and Michael Charles Tobias*

SelectBooks, Inc.
New York

This edition published by SelectBooks, Inc.
For information address SelectBooks, Inc., New York, New York.

First Edition

ISBN 978-1-59079-451-7

Library of Congress Cataloging-in-Publication Data

Names: Laszlo, Ervin, 1932- author. | Tobias, Michael, author.
Title: The Tuscan dialogues : the earth, our future, and the scope of human
  consciousness : a six-day conversation between Ervin Laszlo and Michael
  Charles Tobias.
Description: New York : SelectBooks, Inc., 2019. | Series: New paradigm
books
  of the Laszlo Institute of New Paradigm Research
Identifiers: LCCN 2017049580 | ISBN 9781590794517 (pbk. : alk. paper)
Subjects: LCSH: Nature--Effect of human beings on. | Global environmental
  change. | Biodiversity. | Human ecology.
Classification: LCC GF75 .L38 2019 | DDC 304.2--dc23 LC record available at
  https://lccn.loc.gov/2017049580

Book design by Janice Benight

Manufactured in the United States of America
10 9 8 7 6 5 4 3 2 1

*To our great-grandchildren Houston and Abigail,
that they may carry forward this dialogue wherever they live
on the planet*
ERVIN AND CARITA LASZLO

For my mother, Betty Mae Weber Tobias, a remarkable person
MICHAEL TOBIAS

Contents

Authors' Preface

[I]

Conversations are critical at this ecological threshold in the human story. And Tuscany, in many respects, has long been a pillar of humanity's struggle to test the outermost boundaries of art, consciousness, and the true meaning of a renaissance.

Ervin Laszlo's integration of music, philosophy, physics, and the study of evolution and of consciousness has resulted in remarkable findings, both intuitive and empirical; deeply personal but also universally peer-reviewed. His ontological peregrinations have coupled matter and mind and heart and soul in a dozen disciplines all of which are of keen interest to me because of my own fascination with, and deep immersions within, the scope of human consciousness, particularly the ecological crisis known as the Anthropocene that now rains down upon the world. I have tracked this crisis through intense field research in nearly ninety countries over the course of nearly half a century.

From the first moment we begin discussing it, Laszlo and I ask whether this current ecological fiasco, which is unlike any in the last sixty-five million years and is marked by an absolute loss of biodiversity as well as the potential for the actual extinction of our species, will, in fact, even be noticed by the cosmos.

But within minutes we respectively concur that there is an open-ended road map before all of us yet to be traveled. If cosmology and the study of evolution and consciousness are to have a level of meaning that might find traction amid a large constituency, such discussions must come down to earth. Somehow, the Sun must speak with the sunflower; the galaxies must provide a window of insights that are a wake-up call on this momentous cusp of a terrifying bifurcation—what Laszlo has eloquently described as "a breakdown or breakthrough." I have characterized this as "a wake-up call here and now, on Earth, in our hearts, our minds; at the dinner table; in the broadest possible swath across that mundane terrain of the everyday."

—MICHAEL CHARLES TOBIAS

[II]

Michael has described the hopes and expectations he attaches to our six-day dialogue. All I need to add is that I welcomed having it take place in the venue I freely chose as my earthly abode—because when I left the UN and decided to take a one-year "sabbatical" prior to returning to my university, this is where I decided to come. I came not for reaping any immediate professional or practical benefit, but because this place inspired in me the kind of thinking and feeling that I sought to evolve. That was more than thirty years ago, and neither I nor my better half, Carita, ever regretted it.

Of course, it gave me pleasure that our conversation was to take place at my home in Tuscany. But when Michael first broached the subject of six-days of conversation I thought he was overshooting the mark. I am used to saying what I have in mind in much less time than that: I seldom engage in a dialogue with friends for much more than an hour. One day is my usual limit for a webinar. So, six days? Michael convinced me that the time was needed for our ambitious project, and I decided to go with it.

This was the second experience—or I should say experiment— of this kind that I have entered into, the first having been a conversation on the evolution of consciousness with Peter Russell and Stanislav Grof. That dialogue, transcribed in a slim volume, met with surprising interest, and was translated into several languages and cited for years afterwards. Yet it took only three days (two and a half to be exact) to complete it. Our Tuscan dialogues have better preconditions with six days, and with a venue that in its own way is as attractive and inspiring as the famous Marine County in California where the Russell/ Grof dialogue took place. For this Tuscany conversation I could not have had a better dialogue partner than Michael Tobias, with his keen mind, wide-ranging knowledge, and deep concern for the future of humankind and of all things on this planet.

Let us see what develops, I thought. So I dipped into the sea of ideas, hunches, concerns, and intuitions that Michael has evoked, and that we then pursued in the embrace of these Tuscan hills.

The mapping of our dialogue in the form of a book was not intended to be a finished literary product; it was meant to be a catalyst that would inspire others to come forward with their own thoughts, hunches, hopes, and intuitions. This is the real objective of our Tuscan experiment. Achieving it does not depend on Michael, myself, or our publisher—it depends on the reader. He or she needs to join the experiment not as a passive bystander, but as an active participant. Today, in the networked world of the internet, this can be readily achieved. Achieving it could have a thought-provoking, even consciousness-changing effect, an effect that is much needed in our crucial times.

I should add that I think in terms of process, and not of product. The conversation reproduced on these pages is not a *product*, but the record of a process of living reflection and exploration. We hope that this process merits to be read, and merits also to be continued. We sincerely hope that it will be. We hope to carry it forward ourselves. Other venues and means of communication can also be found for setting it forth: Michael and I, Tuscany, and our publisher do not have a monopoly on creating and communicating thought- and consciousness-changing ideas.

Evolving our thinking and our consciousness is a precondition of creating a better world, and dialoguing on the issues that confront us is a key to evolving our thinking and our consciousness. Let the conversation begin. Michael and I are ready to talk to each other, and to all who wish to join us, wherever they may be. Tuscany conversations are not limited to Tuscany and to this day—they are "nonlocal" and timeless. Let us make the best use of them, in our and the human family's best interest.

—Ervin Laszlo

DAY ONE

Morning, at Villa Franatoni in the Study

TOBIAS: Ervin, we're at your home in Tuscany. How is it that you chose this spot? Etruscan landscapes, passions, history, and ideas—it seems to be all here.

LASZLO: I am sure you know the expression *spiritus loci*, spirit associated with a place. However, I don't see myself as having lived here in any previous existence. But I do feel that there have been many generations who have lived and worshiped here on this hill where I live now. It's a place which has an air of having been a site of veneration and worship for a very long time and it's something that attracts. I think that the spirit of the human who inhabits the place, especially when generations and generations have lived there, gets imprinted.

So in my view, I didn't choose this place, it chose me. I happened to glance at it, and I don't know to this day why, I wanted to go up from the road below to look at it. After visiting it, my wife and I went to the local coffee bar in the nearby village, Montescudaio, which is just two miles from here, and asked if they knew who was the owner of the place and this section of the hill. They said "that man over there," and pointed out a man in the bar. So I went over and we talked for a while. It turned out that he was totally amazed by our interest in the place, as it had been abandoned for a long period of time. I told him that I was in a hurry because the following day I had to return to my university in America, stopping first to see my parents in Switzerland.

To make a long story short, the owner was surprised that we were interested as no foreigner had shown interest in this property near a sleepy little village that produced olives and some wine and not much else. He told us that we could have it if we wanted. I asked when and he said next week. I said I didn't have time next week, how about tomorrow. The next day we met with the owner again and signed a document confirming our intent to buy, and made a down payment on the spot. The price was ridiculously low because it was a ruin in fact, and nobody thought anybody would be interested in it. The next day we left.

TOBIAS: Audacious, inspired, but not surprising given the astonishing beauty. It is precisely what Ralph Waldo Emerson recommended when he spoke of merging the wild stream and hospitable human domicile as the ultimate juxtaposition in nature. My wife, Jane Gray Morrison, and I have endeavored for some years now to resurrect that same admixture of the wild and interior consciousness enshrined as domesticity at our home in Santa Fe, New Mexico, with a modest library, and outside a stream coming from the 14,000-foot-plus Sangre de Cristo peaks above. We have some of the most beautiful lighting I have ever seen; light, as Einstein described it, piercing junipers, piñon, Norway spruce, aspen, crabapple, pear, apple, white peach. Roses and verbena. Hollyhocks of every imaginable color. Sleepy grass and bluestem. Sideoats grama and silver beard grass. Endless species for pollinators.

LASZLO: There is that merging here, in the historic valley of Cecina— inhabited and even worked continuously for five thousand years. I was determined to come to inhabit the place as soon as possible, so before leaving I contacted somebody who would come and put it into shape sufficiently so that my family and I could come back and live in it. At the time there was no power, no telephone, no water—it was an abandoned stone house that had been built as a chapel. By the time we came back, it was sufficiently in shape for us to spend the summer here. And from then on it was a story of building, constructing, designing, and using our imagination to create, from what was once a chapel, the Farmhouse Franatoni, morphing it into the Casa Franatoni, and then into the Villa Franatoni. And this is where we are now. I suppose it could turn into the Palazzo Franatoni if we really wanted it, but I don't have any intention of doing that.

Tobias: Yes, it seems that Pisa has the monopoly on palazzos, whereas here you have a monopoly on a quality rare in our age: that combination of solitude beside the multitudinous gongs of a nearby church. All this, an ancient nature that is Tuscany: Tuscan history, archaeology, biodiversity, and such pluralistic cultures. But how was it you were even in a position to be able to glance at this place, as you put it, and then make it your own domicile?

Laszlo: Because I love Italy and come here whenever I can.

Tobias: Ahhh. Now you're talking!

Laszlo: We lived in Switzerland and for vacation with my family we always came here. And before that, I used to come give concerts in this country. Some of my earliest concerts were here when I came back from America and was on my own. I think I was nineteen years old. I gave a concert in Livorno, which is not very far from here, for the Amici della Musica, and another concert then in Naples. But then, when I settled down much later in my life and had a family, I lived in Switzerland and I drove, from time to time, down to Italy. As soon as we'd cross the border I would start singing—which is something unusual for me and not something I would want to subject you to. . . . I would be in a high mood, coming here to what the Italians call the *terra de la libertà*, the land of freedom. Here you can act and live as you want to.

Tuscany was the place we came to. There is something about the countryside, something about the timeless hills and the timeless human habitations that become part of the landscape. It's a "human" place. The Mare Nostrum, the Mediterranean, is only about ten miles from here. The birthplace of Western Civilization. It's all here. I always liked to be here. My wanting to come here and be here was not much verbalized, not even recognized, but instinctive. I feel myself, somehow one degree more alive here than anywhere else.

TOBIAS: We were discussing last night the notion that there has been at least five thousand years of human habitation here and right there behind us, at the Cecina River Valley down below this hill atop which we are firmly planted. What a view. What a wild, wild river, notwithstanding the fact Tuscany is not exactly the Himalayas. But it is partially wild, in some ways. If you consider the extent of the remaining ecological corridors that are left in the twenty-first century it remains wild. Last night, around midnight, I made my way down the dirt road into a dark thicket beneath the full moon. My god, it was perfect. I was lost, so to speak; no idea where I was, only that I was somehow both vulnerable but safe as a result of all of Tuscany's history. Leonardo, Galileo, and Pinocchio were all looking after me. There was a breeze, in and out of the deep foliage; that pregnant moon— and I was there exposed to everything in the world, it seemed. I felt a distinct collaborative spirit in the air and I was on a meadow and there had been a forest clearance by farmers, of course—olives, wine, wheat, you name it. But the still unchecked river was flowing—out of compliance with both humanity and the biosphere, you might say—not more than fifty feet away. I could see animal prints. Clearly there is a fellowship that encompasses not only human habitations but also wild boar, wild cats, deer, rabbits, polecats, a fox, and the eyes of birds at night, reflected off the moon. Perhaps an owl. An entire pergola of flora and fauna. And you've lived here since?

LASZLO: I have lived here since the mid-1980s.

TOBIAS: So if you were to capture verbally the spirit that has entered your veins, what dominates the sensation from this hill that gives you wonderment, joy, and spiritual and philosophical solace?

LASZLO: A humanness, a human dimension. Something where you can feel that the surroundings are not a stranger, not something you impose on, but rather something that opens its arms to you and into which you can enter. A slight exaggeration maybe, but you can actually feel that you could become one with it. A sense of home away from home. And I should add that I have not ever really had a home. I left my native Hungary when I was fifteen. Since then I have never felt that I had let down roots anywhere. If you ask me where I'm from, I ask you what you mean—where I was born or where I live or where I work? They are all different places. Tuscany isn't my own exclusive place on this earth, it's a place where I feel at home. A place where nothing is forced and artificial. A place that accepts me and doesn't want to make me into something else.

In a way, a home always limits you. It is where everywhere else is a foreign place. When I am in Tuscany, "elsewhere" is not a foreign place, because this is not my exclusive home, it is a place where I can be myself, because nature and people accept me. The Italian culture is a welcoming culture. Italians like people coming from other countries and they don't want to make them into Italians. It is a tolerant culture, a trait that goes back thousands of years.

TOBIAS: I wonder if the condition of feeling at home is something that is unique to our species. In other words, if we were to take a poll worldwide of human beings, how many are likely to say that they feel at home? I suspect the data would reflect not only a moving target but one filled with nostalgia, delusion, denial, and with the sad truth that so many of us are disjointed and lost—perpetual immigrants and refugees. Many of them are climate change, environmental refugees. What do you think?

LASZLO: I wonder if one can altogether feel fully at home in such a rapidly changing world. If the world is changing faster than we can acclimate to it then it is obviously changing under our feet. Maybe young people are already part of this rapidly changing world. I am not one of them. I have not had the problem of not feeling at home because I never felt uniquely and exclusively at home anywhere. But I think that for most people the world is changing too fast, beyond the limits of their acceptance and adaptability. This is why we have so much disharmony, so much alienation, so much strangeness, so many unexpected events and relationships in today's world. I think that very few people truly feel that this world is exactly the way they would like it to be, that it is the place where they feel that they truly belong.

TOBIAS: It's interesting. The concept of feeling at home, *sentirsi a casa, de se sentir à la maison,* and the old Latin, *sentire domi* . . . so much is inherent to the word *feeling*, far more so than the home itself, the languages seem to suggest. Even in your native Hungarian, what does it take for you to feel at home: *hogy otthon erezd magad*. Since I find Hungarian impossibly difficult, that takes it to a new level of mental complexity, at least for me. *Oikia* in Greek, as in *oikologia*, ecology. That begins to tell the truth beneath the languages.

Good housekeeping is the basis for so much of our Arcadian Greek and Roman notions of the Golden Age; of looking back or looking forward to some kind of paradise, and feeling at home in a world where it's safe to say most humans feel alienated. Statistically we know all of us will suffer bouts of depression, often acute levels of anxiety during our lifetime; we have a lot of data that indicate this, though I'm not aware of any such data being correlated expressly with where we live. I suspect that if we were to ask any *Homo sapiens* if he/she felt at home, the answer would probably be, "How can you ask such a silly question? Feeling at home, are you kidding? My children and grandchildren all have problems. Climate change, extinctions, so many issues in the economy. . . . If I'm in Greece I'm worrying over whether Germany will continue to aid us to avert our having to return to the drachma as our preferred medium of exchange. If I'm in Libya or Syria—forget it. I need a refrigerator. My kids need electricity."

But if I ask the sharks, who've been around for 110 million years, they'll tell me perhaps things were going well until recently, until you guys came around. Things were going extremely well for a long time. We haven't had to change, to morph or evolve, we've been at home in these oceans for countless generations. But it is rare to find a human being who can even imagine being at home. That's odd because almost 7.4 billion of us live in houses, most of us in urban clusters— metropolitan areas, of more than 50,000 people. And we know that soon there will be 10,000 cities that exceed 50,000 people. Many of them will exceed a million human inhabitants, some of them even 30 million. And these people are troubled. The countryside is emptying from China to Argentina, people are migrating, looking for something. What are they looking for? You, Ervin, have found this place. What about the people in Rome just a few hours down the road, what do you think, do they feel at home?

LASZLO: People don't have the time to ask themselves these questions, certainly not to think about them. They are taken up with the whirlwind of existence, which means trying to maintain themselves and perhaps their family in a condition in which they can survive. If they are ambitious and also a bit lucky, then they can think about progressing and having a better life for themselves. But being at home for most people is only where you hang your hat, as the saying goes—at least for people living in a city. Yet many people who live in the country want to leave their home because they are looking for a better place to live, somewhere where they can make a decent living.

As a result, people are populating the planet in some unimaginable places. How many habitats are in the world where we wouldn't think that people would or even could live—yet they live there. And they feel at home to some extent. They would feel alienated if they had to live someplace else.

This mobility induced living pattern is a twenty-first century phenomenon. If it stretches the limits of our being able to feel truly at home, it is an unhealthy phenomenon. We are children of the earth and we are built into our habitat on this earth. If we don't recognize that, don't feel that, then we don't have a ground under our feet.

I think the so-called "simple" people—who are probably actually the most sophisticated people in the world—want to live simply. They don't always necessarily want to have the latest frills and gadgets. There are many people like that. I see them everywhere. They want to live where they can see the sky and have contact with nature and with others around them. They still have a sense of being someplace where they can feel at home.

TOBIAS: All right, so let's say we establish a baseline for observing and qualifying and attempting to understand the human condition on the basis of what it takes to feel at home, to be at home—qualitatively more than merely hanging one's hat, as you say; and to ultimately take care of our home so that our children can also enjoy it. That may be difficult in the future when you have a country like Brazil urging parents to forego children, not on ecological grounds, per se, but because of the mosquitos and the resulting brain disease in the children. The question for me is terrifying because I see our home, Gaia, being destroyed.

I acknowledge what you've just said about populations of humans remaining throughout the world, even on the seventh continent of Antarctica where there's now a school with children and teachers. But throughout the world there is the migration, the mobility, the rapidity of change, and the corollaries that come from our consumption as a bipedal and mostly carnivorous species—a recent development, after all, since we're only 330,000 years old or so in this morphological form. When you think about what we have accomplished and also what we have wreaked, all the revelatory power of our joys, our artistic passions, our scientific explorations, our questioning, our curiosity. . . . Well, it has left us in some way marooned as a species.

Sure, we went to the moon and sent a probe to Mars and close to Pluto. But frankly, these ingenious toys do not impress me very much. Even as we sit here comfortably in Tuscany in your beautiful home, in the welcoming culture you've described, we see the challenge our species has imposed on itself, and on all the other species with which we cohabit the planet. We are transgressing all the planetary boundaries that define our ecological home, from overkill of biodiversity, to intense loading of nitrogen and phosphorus, to increasing the global temperature, the loss of forests, the near-loss of the Aral Sea and the total pollution of the Baltic Sea, and so on.

We are a species that has the power to create the *St. Matthew Passion* by Bach, to toss Frisbees, to delight in chocolate chip cookies, but also to have this biology of transgression, that universal Hitler that Jean-Paul Sartre suggested may reside in more than just one person. Even if there were no Jews, there would still be anti-Semitism. These evils seem to accompany our greatest achievements, our greatest joys, our greatest predispositions to love, and our ability for compassion and ethical values. I'm troubled by this apparent dialectic. Being at home in a world where, in essence, I personally feel persistent shame. I feel almost caught in some moral, conceptual, and physical cul-de-sac out of which it's very hard for me to see my way free. I'm bound to my species.

LASZLO: This is certainly one aspect of what is happening. There are other aspects as well. What is happening at the same time is an increase in contact and communication harboring the possibility of creating new relationships. This world is in evolution, changing from fragmentation toward unity. But I can't say that the world is all in fragmentation or that it is all moving toward unity. There is movement toward connection and relationships. This movement can be exploited and misused, and it's often misused. At the same time we can see that underlying the processes and developments in this world there is a new level of contact and communication informed by a new level of consciousness. Sometimes mistaken, sometimes only partial and very often unknown to the individuals who are involved with it, but there is a new form of communication emerging on the planet. It is based on the kind of connectivity inscribed at the heart of nature. Its genuine expression on this planet is overshadowed by a headlong rush toward change, leading us towards a world that we no longer understand, toward one in which we do not feel at home. Yet underneath the rush and the alienation there is something new emerging: a pooling of individual consciousnesses into something like a collective consciousness.

TOBIAS: In any large human population center, I see the impact of religion, of spirituality, of demographics, of the ratio of men to women and their needs and their degree of freedom, or lack thereof, because of their socioeconomic situation; I see that all these ultimately ephemeral considerations inform our emerging collective consciousness. Remember that each of our lives encompasses, if we are lucky, three to four generations, and then the individual is gone, we're dead. Some 108 billion people have come and gone during our presence on the planet. Do you see humanity as a major player in the evolution of this planet or are we simply one more passing vogue in the annals of cosmology? In the annals of biology let us be clear: we are just a passing vogue and all of these considerations are ultimately irrelevant. And as far as the revelations of Carl Linnaeus, Comte de Buffon, Jean-Baptiste Lamarck, and Charles Darwin, none of it matters. And frankly, nothing Einstein said has much relevancy in my mind if our own ecological standing and ability to rely on the notion of a "home" is equally ephemeral, without substance or true meaning.

LASZLO: I believe that there is an underlying evolutionary trend which is towards coherence and complexity; finding expression both in behavior and in thought and spirit. In that sense humanity is advantaged because we have developed the ability to infuse consciousness with our actions, and we are able to estimate and to reassess, if we want to, the past, and are also able to predict, or at least to speculate upon, the future to an extent which is greater than in any other species. It is unique to our species on this earth.

TOBIAS: What if it's not?

LASZLO: What do you mean by that?

TOBIAS: What if the consensus of science and of psychology, of research in general, of anecdotal experience that becomes a science in and of itself; what if we were to turn around and look in the mirror in five years, ten years, tomorrow, and see that the unique qualities that we've attributed to our species in fact have been around long before us, among all living beings. I prefer to be nonexclusive if I'm going to start attributing sentience to nonhumans. We now know for sure it exists among many species including chimpanzees, bonobos, dolphins, probably blue whales, and perhaps octopuses, which is saying something definitive, considering that nearly 70 percent of an octopus's neurons are not in its brain but in its arms.

The topology of what I term the semiosphere (first applied in 1984 by Yuri Lotman) suggests to me a whole new reassessment of life and consciousness that is way overdue. Remember, it was Michel de Montaigne who declared that other species are far more rational than humans. Dendrologists, who study trees, would argue that the redwoods, which have been around for nearly three thousand years, communicate and are geniuses all their own. They might conclude that their very endurance, size, and volume of life is so finely tuned that something must be happening in terms of communication between and within the trees themselves. They can reorient directionality of growth, not unlike your solar array out in the back, moving consistently with the sun. Redwoods actually orient in terms of moisture, in terms of how they are fixated on a gradient to accept the most amount of precipitation on their leaves and in their bark, as one would expect, even though our understanding of photosynthesis suggests that, by our standards, it is very inefficient, at approximately 10 to 15 percent of the theoretical potential in any plant. But the redwood's life span, especially in the *Sequoiadendron giganteum,* is proof positive that what they're accomplishing is rare and magnificent. In addition, they support a remarkable assemblage of life-forms in their upper canopies.

So my point is, if we're celebrating our consciousness as being endowed with some special ability, as you were suggesting, what happens if we discover we aren't special at all? And not just not special, but possibly a bit deformed? I mean, what other species would knowingly destroy the Aral Sea or ignore lead in a Michigan city's drinking water? Or, for that matter, kill most of the indigenous Native Americans in a place called Yosemite just seventeen years before John Muir himself arrived in that emptied out wilderness and called it sublime, like some icon, ignoring the fact that Native Americans in Yosemite had been murdered, bled out, so to speak, like sheep in a slaughter house. What other species would do all this?

Yes, now there are a few dam projects in the Aral Sea to try and restore at least a fraction of the fishing industry. But again: what does it mean? Restore an industry which is anything but restorative? Rather it's renewing the penchant for killing sentient beings. And ultimately, what happens if we discover that our perception is no different than those who are perceiving us among other species? What if we truly discover, as Nietzsche did with his concept of God, that Man, too, is dead?

LASZLO: Then they are just proving the point that you are making. I believe that we can discover or rediscover that humans have an ability that is practically unique at this point of evolution on Earth.

TOBIAS: Well we could use it, we could be empowered by it, certainly enlightened by it. But I do wonder if it would help us out of our logjam of emotional and cognitive stubbornness? Most people find it difficult to relinquish the idea of their personal superiority.

LASZLO: In my opinion, we have to distinguish between potential and actualization of the potential and see how we can assess the various forms of life and expressions of life, ranging them along a spectrum that goes from the built-in potential, which is there in the single-celled organism, in the primordial hydrogen atom, to its fuller expression and unfolding. Is this potential, which we call mind or consciousness, really a side effect? An epiphenomenon? Is consciousness something produced on the side, in the course of producing something else? Or is consciousness an expression of something fundamental, a potential residing in the cosmos?

This is the basic question. It's a core question. How do you look at the world? As something which is just one darn thing after the other? I think we have now seen that the probabilities of a universe even remotely ordered such as ours coming about via random processes is astronomically small. So there is something happening that gives a vector, a direction, a directionality of change. Could it be that this directionality is the consciousness element?

TOBIAS: A multiplicity of ways of assessing this question, which might all be right, or all be wrong. Remember Shakespeare's *Hamlet*:

> And therefore as a stranger give it welcome.
> There are more things in heaven and earth, Horatio,
> Than are dreamt of in your philosophy.
>
> (*HAMLET* ACT 1, SCENE 5, 159–167)

LASZLO: We can be either a reductionist or an idealist. Or take many other positions in between. But there is something that could be called objective reality, something that appears as what I call coherent complexity. More and more structures emerge that are improbable in terms of pure chance. Something is emerging that has coherence, has structure, and has an element we describe as self-recognition. We don't know how often this is happening in the universe. There is a possibility opening for a whole range of speculation, for example that evolution is leading toward a kind of unfolding of an intellectual or spiritual potential in the universe and not only on Earth. I don't think that we are unique. But there certainly is an element that could be an expression of a core trend in the cosmos. We can see consciousness like this, or we can see it as a meaningless by-product. This is a choice we have. We have to choose. But today the choice, I think, is made more on the basis of intuition than reason.

TOBIAS: You think we have the freedom of choice?

LASZLO: Of conceptualizing what we see as the real world, yes.

TOBIAS: And beyond that? In terms of action?

LASZLO: There is no absolute freedom in the universe, because there are laws that constrain the behavior and the evolution of all things. But these are not deterministic laws but laws of probability.

TOBIAS: So if we take James Lovelock's Gaia hypothesis, for example, or embrace the notion that the earth herself, as the Greeks so ordained, has a purpose—forget God, forget atheism, forget faith or any religious orientation—but a biological predilection that is evolving toward some self-expression, some self-consciousness, some global consciousness that is oriented and predisposed to create life on earth, that is mere probability? Like winning the lottery or drawing a full house in Vegas? How can we call it probability in the face of other star systems and the relatively recent revolution in astronomy that tells us about thousands of other planets when just a generation ago there was skepticism about any other planets let alone the recently discovered eight planets capable of bearing life in the so-called Goldilocks zone?

We've come a long way in very quick blink of the eye. But if the Gaia hypothesis, for example, in which it is predicated that life itself and the whole planet is a living organism, were true and we're a member of that community, what is the worth of our consciousness unless one takes what I would deem an egotistical position and suggest, as remarked upon earlier in our conversations, a kind of superiority factor? This pathway I personally reject because I see us as members who go to communion, if you will, every time we breathe in oxygen, every time we look at the night sky, every time we take delight in the wind against our cheek as Shakespeare described in *King Lear*.

I'm one who immerses himself in the forest, in the sea, in every possible biome because I feel humble before this vast assemblage of creatures looking at me, breathing with me, and I am thusly and gratefully minimized by my own enthrallment with nature that I scarcely have the audacity, if you will, to wonder whether what I have to say or think has the slightest importance. Is the idea in the mind of a flea of interest to evolution or to Gaia herself? I'm an ant, a distressed and at times satisfied ant, but hardly important. Unless, of course, one takes the enlightened view that every flea is miraculous and weighs heavily on evolution, however unseen its effect. Miriam Rothschild certainly thought so. I've devoured parts of her six-volume work on fleas which she wrote with George Henry Evans Hopkins between 1953 and 1981.

LASZLO: You are assessing importance in terms of action and of spontaneous recognition. You look at it in terms of self-cognized elements that lead you to act in a certain way and then ask whether that action could have an impact or influence on others. But consider that the rational recognition of things within the overall trend may be secondary to one's participation in that trend.

TOBIAS: Ervin, we're at a juncture it seems to me. Humanity is at a turning point that, in my view, is more serious, exciting, challenging, and terrifying than any in its history. As I said a few minutes ago, clearly there is a dialectic. We have this dual penchant as a species, and if history is any record, and it is I think, we have demonstrated both violent and nonviolent tempers, love and hate, and dualities in general. One of the great books on symbolism that truly underscores this schizoid behavioral tendency is Simon Schama's *Landscape and Memory*. Schama describes, for example, how by the time of the Norman conquest of England in 1066, some 85 percent of the forests of England had already been chopped down and how it continued to get worse. And here, not that far from your backyard—aside from approximately two hundred protected areas across Italy and the high degree of forest cover throughout the region—a similar trend can be deduced.

From France to Russia, we've seen similar decimations throughout history. It started thousands of years ago with megafaunal extinctions perpetrated solely by human beings who took advantage of the retreating ice fields during the fourth glaciation. Opportunism throughout human history is mostly violent. We are almost recklessly capable, it would seem, of nearly anything. If we can't fly like a bird we'll build an airplane; if we can't run as fast as a horse we'll build a Porsche; if we can't swim as seamlessly as a shark we'll build giant cruise ships with five thousand people on a boat. If we look up at the moon with wonder, we decide to go there.

Now we're talking about going to Mars. Just the other day the latest Buzz Aldrin plan was announced followed by the Elon Musk's announcement regarding Mars—a new civilization, terraforming, and tour groups to Mars and the moon. Hurrah! We seem to be a nomadic species. We want to see over the hill and see what's on the other side, which is a wonderful quality. Almost every young person is anxious to indulge his or her curiosity and go out there.

My mother, who is ninety-one years old, says, "Well, Son, my time is nearly up. It's your turn, although even for your generation it's probably too late. It's going to be the fifteen-year-olds, the ten-year-olds, the eight-year-olds." I heard that a five-year-old who made some YouTube video got over a billion hits in 2015—the most watched video of all time. What does that say? I'm not sure but the world we're leaving is for them. And as my mother intimates, is their world going to be significantly different than your world and my world? And she says absolutely so. I had this discussion with her just a few nights ago and she reminded me that she lived through World War II, she saw what the Nazis did, and what the Americans did to Hiroshima and Nagasaki, and what the Japanese did at Pearl Harbor. She lived through it all. We've learned from that, supposedly. But have we? Are we really improving as a species? Could the US and North Korea engage in the moronic, sadistic globally unthinkable? And does it even matter one way or the other, in the broader view of things?

My question is, when you talk about core, beyond simply improving ourselves as a species, what is the real challenge before us? Because improving as a species is solipsistic, narcissistic, it's ego driven, it's about us. Is there something greater than us?

DAY ONE

—◆—

Afternoon, on the Front Patio

LASZLO: You spoke about improving ourselves as a species and called it ego-driven. Let me put this in context. To improve is to do, create, or act better than one did before. For this we need to know what is "before"—what is the state or condition of the world. It's a big question, and to answer it we must first see at what level of resolution we assess that state or condition. You are looking at the world at a medium level of resolution, in between the quantum level and the planetary level. You are looking at the level of an ecologist and you see this tremendous turmoil in human societies which itself is an indication of change at an accelerating rate. We can choose to look at the world at different levels of resolution, for example, the level of the quantum.

Why is it that we don't fully know what brought about the world? There are some physicists who contest that the Big Bang happened, but in general there is a consensus that something fundamental happened 13.8 billion years ago. Why is it that at that point in time a sufficient number of particles survived the initial collisions to form stable orbits around nuclei with positive and negatively charged particles? Why didn't the particles just agglomerate around nuclei, and why is it that no two electrons turned out to occupy the same energy level, with the same momentum, in the same place with the same position? Physicists explain this in reference to the Pauli exclusion principle. First particles emerged out of what appears to be an underlying sea of energy. Why did they emerge? And why did they start building up? Without a process based on this principle, there would not be atoms, atomic structures, molecules, supermolecules, life and life on planets, or even planets and stars.

So there is something from that singularity onwards which is building structure, building it continuously but not in a linear way, with a lot of fall backs, a lot of fluctuations. Life is an expression of this buildup of structure. Basically, all structures are emerging from the same cosmic ground. And all are emerging in the direction of increasing complexity with an astonishing level of coherence. Einstein remarked that the most remarkable fact about the universe is that it is coherent and that we can grasp it with coherent reasoning. The universe is not accidental. And at the same time, it is globally coherent.

However, we can locally subvert the global directionality of the process. We can create processes that lead into a false direction; we can become a part of the process that dies out. Human life on Earth is a branch that would die out, because it has deviated from the unfolding of the basic process and endangers other branches around it. We humans are now capable of destroying ourselves and all life on Earth. Given our technologies, that is possible.

Evolution in nature is a surge, a seeking, a general direction in which it exploits all possibilities. It's relentless, it's unceasing, it seeks whatever possibilities there are for life—as you well know as an ecologist. Wherever there is a possible habitat on this planet some form of life, some species will occupy it and use it and develop in and with it. All evolution in the universe is of this kind. On this planet we are in a buildup phase right now and we don't know whether this buildup phase will reverse into a phase in which the cosmic average temperatures will heat up and ultimately nothing which we recognize as life will survive.

Is the universe open or closed, as astronomers ask. We don't know for sure, but we do know that in either case life could not exist forever. But evolution could be cyclic, meaning that this phase of the universal buildup could produce forms of life that evolve again and again from the ground from which space-time itself emerged. Then the next universe, or the next cycle of this universe, would build on these forms of life.

TOBIAS: When you say the "next universe," you're saying extragalactic—outside of this planet or what? We know that this planet will be gone in four billion years.

LASZLO: I mean a process of evolutionary buildup beyond this part of the universe. The universe has far more galaxies and planetary systems within the galaxies than we thought. A hundred years ago we thought that our Milky Way galaxy was the whole universe. In the emerging concept the universe is the domain or the territory that is reached by light since the birth of light and of all things some 13.8 billion years ago. The periphery of that territory is now about 42 billion light years from Earth.

TOBIAS: This question reminds me of the whole field of topophilia, where perspectives shift dramatically from local to nonlocal experience in regard to ethnography, as Yi Fu Tuan and others have written for decades. Yi Fu Tuan is particularly interesting when you look at his broad canvas as a geographer with the breadth, one might suggest, of the German painter Anselm Kiefer (particularly the Kiefer that Simon Schama discusses in *Landscape and Memory*, with a chilling early German Green movement translation of the Holocaust and the destruction of forests into a thesis that goes to the core of human nature). In the case of Yi Fu Tuan, his books on human goodness and the landscape, climate, cosmology, in addition to a work published by the University of Toronto in 1968 entitled *The Hydrological Cycle and the Wisdom of God*, are very interesting, indeed.

You see, I keep wanting to bring down the large philosophical umbrella of the universe and try somehow to both grasp and make relevant for ourselves the preconditions for survival here on this tiny planet. And in doing so I want to raise those specters of change and of potential, as one might do with electricity, but instead by implementing the human spirit—how we can be champions and torchbearers of biology. Because we've seen, in relatively recent times, the challenges to our existence. Seventy thousand years ago the Toba supervolcano erupted in what is today Sumatra and nearly destroyed our entire species. We were left with an estimated twenty thousand breeding pairs in what can be described as a truly unprecedented genetic squeeze. And yet, from a global population of perhaps forty thousand, we continued to flower, to make more than a handsome comeback. We returned in a fury, with a vengeance. And now, of course, we're headed towards ten, eleven, twelve billion people by the end of 2100. If marabunta, the nearly two hundred species of army ants who have most likely exploited what we term convergent evolution, were our size, can you imagine the world? Well, there may be trillions of ants, and each one of them imagines a world which to us is frankly unimaginable, but they see and feel us in ways we will never understand—we are enormous creatures right out of science fiction—and they certainly see what we're doing to the Amazon and every other habitat they utilize.

The UN, the IMF, and the World Bank keep revising the medium numbers of how many we are, and by consumer implication where we're going. But we are unlike those army ants, who will stop at nothing to meet their dietary desires, in that they are small and their impact totally sustainable within whatever environment they inhabit.

We, on the other hand, are huge vertebrates, relatively speaking. And we are continuing to expand our hegemony across the biosphere in every way on every one of the seven continents. We do so with such reckless arrogance and youthful enthusiasm because we're so enamored of our toys, of our power, and of the corruption that power inevitably brings. The great historian Arnold Toynbee said this with respect to the twenty-two known past civilizations that have gone extinct, like the Aztecs who brought themselves down through their arrogance. I would describe it as inherent evil when you leave a legacy of something like three million young people who were eviscerated atop sacrificial altars to invisible gods, their hearts ripped from their still living bodies. But it's a mindset, a prevailing bias of a culture. The differences are so stark within and without. Consider, for example, the vast extremes that Saint Luke depicted between Jesus and Mary in the manger and Caesar Augustus. Worlds apart. Or the madmen Hermann Göring and Heinrich Himmler standing proudly amid the ancient forest of Bialowieza, determined to see it become Teutonic with all of its non-German inhabitants annihilated. And I see it again now in our meting out of injustice, of unfairness, of outright cruelty to ourselves and to life around us, by a seeming few. This is not something ants have ever imagined. I am certain of that.

Now, if we grant that life around us is merely one phase, a biological cosmogenesis if you will, one aspect within this greater billions of years of a biological umbrella, my mother is not going to be interested in that. She'll say, "I don't care about space, I don't care about cells." (She does, of course, if she goes to the doctor, she, like everybody else worries perpetually about a cancer diagnosis, and she hopes her doctor knows what he's doing and is board-certified.) But in the short term we have very small horizons. So, yes the great philosophical questions that we ask as a species—and have since we could record those questions, whether as petroglyphs, or in a Plato's *Republic*—are meaningful to us; with the poignancy of Ottorino Respighi's *Ancient Airs and Dances*.

Respighi was fortunate indeed to die three years before the Germans invaded Poland. He had remained neutral regarding Mussolini, and, as you may know, was a close friend until his death of Enrico Fermi. They had an ongoing dialogue, evidently about the differences of music and physics. Fermi was convinced *Ancient Airs and Dances* could be explained in terms of physics. Respighi was of an entirely different mindset. They never agreed on a common language. It would be like trying to translate the fusion of consciousness with the universe in a language of mathematics as opposed to, say, in the language of the great Sufi poet of the twelfth and thirteenth century, Attar of Nishapur, whose epic, *The Conference of Birds* is simply beyond powers of ten, or any mathematical or plain-speaking astro-geophysical translation. It is human wisdom beyond the powers of reductionism, in my opinion.

My sense is that you're saying all human affairs are not just meaningful to us, but rather mark some kind of meaningful moment in a much bigger evolution that encompasses some dazzling truth far greater than the earth and her multitude of living inhabitants—far greater than the word biology and all that it implies—something that touches upon a conscious evolution that is beyond description. It is beyond that *something* of which Max Planck said that you cannot get behind, because when you are asking the question it is still beyond what you are asking. How important is that?

Laszlo: It's important to recognize that our understanding encompasses only a very small segment of the world, and that what is happening in the world is happening whether or not we understand it. Our behavior can make a difference to it at some critical points, and at a crucial point a so-called bifurcation point it can make a major difference. But our understanding of the world is ordinarily not the crucial factor.

Tobias: What is crucial?

Laszlo: Understanding that it is happening whether or not we understand that it is happening, or why it is happening. It is crucial to understand that what is happening is governed and decided by information. Information is basic to the world. This cosmically decisive information is what physicists describe as *superinformation*. This is the information that is expressed in and governs all things in the universe. This is not a random universe; it has a basic, built-in tendency toward coherence. We are a minute element of the process that manifests this tendency, but we can be a crucial element because we recognize it and express it explicitly. What counts is the recognition that there is a basic superinformation guiding processes in the universe. We can call this a divine orientation, or we can ascribe it to the laws of nature.

TOBIAS: Do explain, please.

LASZLO: These laws are the fundamental regularities, the intelligence behind matter. For this we can also use a traditional expression, calling it divine providence. Fundamentally the laws of nature express an intelligence because the universe is coherent and what happens in it is not accidental. There is a buildup of atomic structures that creates molecular structures, which create cellular structures and biological and then macrodimensional cosmological structures. Something is building up in the universe in a nonrandom way. The question is, can we recognize it—and can we align ourself with it? The answer is yes, because the orientation, the superinformation, that guides the processes is not outside, but inside us.

TOBIAS: That strikes a chord in me that I feel very motivated by. Namely, going back to the dialectic that I referenced—the polarities in the human experience—and also going back to the beginning of this conversation about your home. Now we have the inside and we have the outside. We have the indwelling, the inner being, and the outer experience. It's fundamental to every science; it's fundamental to medicine, to all inquiry, to thought, to reflection, to apperception—the perception of perception. The question that devolves at this moment in my mind is this: If I look historically back at human willpower, at faith, or anthropologically back at the evolution of art, as manifested in so many relics and documents, or in what you have described in great detail throughout your career as the Akashic records, there is something there which Planck knew, even if he refused to acknowledge our ability to fully grasp it.

We've seen it underlying civilization after civilization, from the Rosetta stone to *The Marriage of Figaro*. Call it divine, call it a creative impulse in the human species to go outward and then bring that experience deeply inward and churn it up, as within Hindu mythology, with a big butter churner. Then we come out with some expression that you could really argue as a cosmologist is an expression of the planet, of the galaxies; it's really some minute expression of the star system that we happen to inhabit at this moment in time. Those are laws of nature, it would seem to me.

But it also appears that we are somehow grappling with the inner and the outer. I used the word "marooned" earlier to describe my own personal feelings about this. And it would help me to better understand in your quest, throughout your life, how you have come to be so impassioned and willful and deeply challenged by the whole notion of human consciousness. Why it matters to you to the extent it does. You use that word "consciousness" very frequently. Why is that word so fundamental to your quest?

LASZLO: I'm glad you ask me about consciousness, because what challenges me, what interests me deeply, is the cosmic consciousness of which human consciousness is one particular expression.

Consciousness in the world is not limited to human consciousness; to the brain and body of a single species. It is a fundamental element of the reality of what the Greeks called *kosmos*. In that sense consciousness is the intellect, the Logos, the motivating, guiding, orienting factor that orients the universe from its birth through sequentially arising and collapsing multiple universes.

We have good reason to suspect that this is the case, and also that through this sequential arising and collapsing something fundamental is unfolding. What could that be? The materialist-reductionist observer says what is unfolding is different configurations in the motion of matter. Matter is becoming more and more structured. And then the time comes when the structures are breaking down and "matter" is restructured again in a constantly iterating process. But when we recognize, as Max Planck did, that there is no such thing as matter in the universe, what we find as fundamental in the universe is a vast information and energy field creating pattern after pattern, resonating and creating structure after structure.

And if that is what is happening, we can ask, why is it happening?

The world is not a blind concourse of bits of matter, as Newtonian classical physics told us. Something is happening that brings more and more frequencies and waves in phase in the universe. We see that when we observe that there are more and more complex and coherent forms of waves and frequencies coming about. If that has a meaning beyond it just happening, then the meaning must be that it enables the emerging clusters of waves and frequencies to evolve a higher and higher form of apprehension, apprehending themselves, apprehending others, and ultimately apprehending the world. The evolution of apprehension in clusters of complex and coherent waves and frequencies could be the ultimate reason for the existence of things.

This possibility is why I am so interested in the presence of consciousness in the universe. We can call it mind, spirit, or soul, we can have many different names for it, every culture has its own name, but they all mean the same thing. They refer to the presence of a super-cosmic intelligence in the universe.

TOBIAS: These are big words.

LASZLO: In the final count what is happening is a nonrandom coherence-oriented process. We are, at least in this corner of the universe, one of its most explicit expressions. An expression that is capable of apprehending it, of participating in it, and yes, even of bringing it forward.

TOBIAS: And yet, in what can only be described as modern times, at the very basis of Darwinism, of the collaborative root of the work of both Darwin and Wallace, random mutations were key to our tenderness, to our being, in this form, to our evolution. Randomness. Darwin had no patience for purpose; he had, from everything I've learned, little patience for the mystery you're evoking.

LASZLO: But that's the point I'm making. We are clear that this is an incorrect explanation. Randomness could never produce even the simplest single-cell organism. The time span of the universe since the Big Bang was too short to produce the level of order we find even in a fruit fly.

TOBIAS: What if Darwin were sitting here and he said to you, "Ervin, random mutations are the only way statistically we can possibly account for the intermingling of genes and the discrete populations that can reproduce themselves."

LASZLO: But we can't account for the order we find by random mutations alone. Fred Hoyle said it beautifully: the probability of a living organism coming about by random mutation is similar to the probability that a hurricane blowing through a scrap yard assembles a working airplane.

We're talking about probabilities, but the probability against randomness in the universe is astronomical. There is no real randomness at all, because nothing is disconnected from the rest. There is only a limited range of probabilities that given things occupy particular spaces at particular times.

TOBIAS: This could be perceived very optimistically. Let me cite, for example, the work of one whom I admire, the organic chemist and molecular biologist Alexander Graham Cairns-Smith, who came up with the crystal theory of evolution. In his theory he suggested that kaolinite—a soft white clay mineral and offspring, if you will, of granite like feldspar and quartzite, the mineral substance of which you find in certain clays all the way from Australia to the Karakoram in Pakistan—produces a microscopic structure that is absolutely the shape of DNA and provided the original sanctuary for rogue DNA that hadn't become life yet. In other words it occupied a niche by providing replication and information processing capacities approximately 3.8 to 4 billion years ago at Shark Bay in Australia, as well as at known sites in Greenland, particularly in the southwest at the Isua Greenstone Belt. This crystal structure in the kaolinite enabled life to advance and utilize a microspace in an informed way and develop polypeptides, amino acids—all the preconditions for prokaryotic single cellular life. This, in turn, would exploit those gained advantages, and by 800 million years ago, eukaryotic multicellular life was flourishing, and by 600 million years ago, the emergence of gymnosperms and angiosperms—the Precambrian/Ediacaran and Cambrian Big Bangs of biodiversity. The rest is biological history.

In other words, this was an evolutionary journey defined by the exploitation of ecological opportunities in ways that took raw matter and converted it to homelands, to territories that could be occupied by life. These were the first homes and homesteads, if you will, which translates into a purposeful, biological odyssey; something on the way to something else that becomes increasingly beautiful, layer by layer more and more harmonious. And if you are telling me, which you are emphatically, that the whole cosmos has replicated in that way and has engendered the preconditions for increasing complexity, then that theory is ultimately not complicated. The cohesiveness and coherence you've referred to—I can appreciate that when I think of music. And I know that having grown up as a classical pianist you would know the minute you heard the slightest atonal instant, an instantaneous off-note, it would send shivers down your spine.

So harmony is something in our blood, something we detect instantly. It's hardwired. Maybe Darwin's and Wallace's mistake was in thinking that in random mutations you get the same preconditions for coherence that enables evolution. Or maybe they were saying the same thing. They had their own mathematical deficiencies, then. They didn't have the Polymerase Chain Reaction (PCR), and they didn't have the ability to get genetic markers in three minutes. They had to go a whole lifetime of studying the shape of mollusks—well, eight years in the case of Darwin. Then he spent twenty years or so on his work with finches in the Galapagos, four species of which we know have undergone rapid adaptive radiation, and the songbirds of England. All towards an effort that would enable Darwin to decipher how discrete populations were essential for evolution to move forward. He had no way of reconciling variations in color and behavior and reproductive success and failure except by acknowledging the existence and nature of those discrete populations.

You are suggesting, as I understand it, that atomic structures—the subatomic particle revolution that had emerged since the wave descriptions in the 1920s and 1930s—have the same evolutionary potential and, in fact, actualizing capacity, as we do. And also that there is a palpable relationship between the Amazon and what is happening in the dark background matter—that there is a direct cause and effect relationship between what's happening in a G2V dwarf star a million light years from here and the famed Carrara marble found in Tuscany. That these relationships are not accidental. That there is a purpose behind all this and we can take heart perhaps in knowing that we are a part of this extraordinary symphony.

LASZLO: That is assuming that the presence of purpose is but an interpretation. If we can say that a process is not random but based on information, then we can conceive that this information expresses something we can perceive as purpose. But purpose can only be expressed by consciousness, a conscious entity of some kind. That is the end towards which the process is moving. Without consciousness, we cannot speak of purpose. We cannot speak of something that is nonaccidentally organized, something that has its own sense of being and reason for being.

TOBIAS: How would you characterize the self in the concept of the self-organizing universe?

LASZLO: Self is every entity and every group of entities that has a coherent core of common features.

TOBIAS: Go on.

LASZLO: A self-resonating in-phase indicates a local self—a local cluster of wave and frequency. Some clusters appear as material objects and some as elements and items of mind and consciousness. Clusters of vibrating energy are the only things we can find at the foundation of reality. These clusters are not accidental—they have a purpose underlying their existence. And that purpose is not a part of what is happening, it is beyond it. It's like chess. A game of chess follows the rules of chess, but the rules are not part of the game. They are on another level, above and beyond the moves that unfold in the game.

TOBIAS: Do you believe in God?

LASZLO: I believe in the intelligence of the cosmos, in a divine intelligence. If you say "God" that means personifying the intelligence. I don't believe that we need to personify it. God could have created us in his image, but we don't need to create God in our image. I believe that there is a reason, there is an intelligence in the cosmos, which means there is likely to be a consciousness. A fascinating and important part to the concerns you're expressing is that the purpose is expressed as information and that information is part of our body, part of our ecology, part of the web of life. It is the same information in different forms.

Today there is something we call alternative medicine—I would rather say complementary medicine—which is in fact an information medicine. It is built on the recognition that there is an ideal pattern of information in our body. In the East it is called the chi and recognized as life's element. It ensures coherence and enables the body to function in harmony, since the body can only function at all if every element of the trillions of cells (more than the number of stars in our galaxy) act coherently. Life is only possible if every element is fine-tuned to every other element and the whole system is tuned towards maintaining the system in its environment. This calls for encompassing and effective information. That information is absolutely basic for life, for the individual, for the ecology, and for the whole planetary system. That information is there in the cosmos, and it is the same information that is expressed in the interaction of atomic structures and in the evolution of galaxies. It is the same information that is expressed in the evolution of life on Earth. It's up to us to recognize it. Because if we recognize it, we could find our way beyond the crisis in which we now find ourselves.

TOBIAS: It's interesting that in Jain tradition, it's fundamental to their belief that there is no beginning and no end in the universe. And they have, after all, been cosmologically driven for many thousands of years. They consider themselves to be cosmologists and have represented pictorially the cosmology of the universe as bound to the twenty-four tirthankaras, or spiritual teachers, like Mahavira, who is the most recent of the twenty-four and a contemporary of Buddha. Mahavira acknowledged that there is no beginning or end, yet human beings have a paramount role to play in Jainism. This was both acknowledged and inherited by various schools of Buddhism. That interests me because regardless of this greater consciousness in the cosmos, we are who we are at a time and in a place that we have to deal with, for better or worse. And we have to deal with our inner nature, too.

For me, the cosmologic wonders and data that are pouring in— the new star systems, potentially habitable planets, and the many moons of these newly discovered planets that we're now able to photograph with unbelievable clarity (which ten years from now will look murky, no doubt)—make the role that I can play as an individual seem all but muted. Thoreau famously suggested that he never felt lonely, not with all the stars overhead. And yet, as you said earlier, we're so troubled, pressured, and challenged just by the rudimentary requirements of our survival that we don't have the luxury to engage in a discussion about much of the data that is at our fingertips. We just haven't rallied with sufficient trust or enthusiasm to get at it, although hundreds of thousands of researchers in hundreds of disciplines are trying.

But even if we could engage at the highest levels, the majority of us are left just trying to get through the day. So I guess the question is, how do we even set up a table of contents for human behavior that gives us the basic orienting mandates and goals and titles, headings, and headlines by which we should be operating to get it right? And that's presupposing that there is a getting it right. How do we set up a cosmic operating system or standard operating procedure that includes humans? It may well be that no matter what we say, think, do, or feel, we're going to get it wrong. It may well be, as you intimated earlier, that we're a tangent, that life on Earth is simply a meaningless by-product of a much greater happening going on in the universe, and that everything we're occupying ourselves with, all of these questions, all of these beautiful dreams (and horrifying nightmares) that we're expressing, are not relevant, except to us.

DAY TWO

Morning, at the Upstairs Terrace

Laszlo: On thinking back about what you were saying, I think you are asking not only about the "how" we are doing what we are doing, but the "what"—just what is it that we are doing.

Tobias: Yes, I am.

Laszlo: I would like first to say something about the "how." The "how" means a series of algorithms, instructions, recipes, commandments, something that directs the workings of an organization; a living organism, indeed of any complex system. Focusing on the "how" is to adopt the manipulative approach. It contrasts with the aesthetic, the poetic approach exemplified by Gandhi's saying about being the change we want in the world. In this approach, we don't intend to change others, we become the change ourselves. We become the correct, the right change when we are aligned with the evolutionary process unfolding around us. Then we are acting for the greater good. Not necessarily for the survival of every part, because the greater good means that certain things or beings need to transform, and some may even have to become extinct to give way to the evolution that goes on around them.

Tobias: We know that. We know that 99 percent of all species have gone extinct.

LASZLO: We are hoping that we can create a perfectly harmonious world, an ideal condition in which every person, every part expresses the wholeness of the entire system. What is good for the entire system is good for every part. That is a theoretical possibility. The "how" is to recognize that we have this information in every cell of our body, and it's this information that runs the Gaia system, which runs our galaxy and the metagalaxy. To access this information is not a question of coming across a blueprint or set of instructions, but of appreciating in a deeper sense of the word what exists and how things are, and then going with it. In other words, it's a seeking of harmony, even if you don't understand what that means in practice. By seeking harmony, we acquire enough information instinctively, to find our way.

TOBIAS: Find our way around our neighborhood which is the biosphere.

LASZLO: Which is this earth.

TOBIAS: Of course. We're not going to be able to impede or delay a supernova. We might be able to terraform Mars but it's unlikely. And going far beyond Mars? Well, it's entirely unlikely unless we reimagine time travel. The technology at our disposal for such bold and probably ill-advised purposes is nowhere close to meeting the mark. Moreover, such initiatives as terraforming strike us as pure hubris. And then there's the whole scenario of leaving Earth worse off than it was before our species existed. That, to my way of thinking, is a desperate epitaph, both painful and deeply frustrating to me as an individual. Given what we believe to be the case regarding the impediments imposed by special relativity, even now we understand that there are vast constraints to our emigration from Earth elsewhere in our solar system. But going beyond our solar system, getting outside the Milky Way, no matter how hard Elon Musk and Richard Branson yearn with the privatization of rocket ships, we are going to have discover and then learn how to safely exploit those legendary, so-called wormholes in order to vastly speed up travel outside the solar system. Which also begs the question: When we are so determined, by all appearances, to destroy life on Earth, why would we ever think to encourage travel beyond Earth, where our human nature would simply perpetuate its tendencies, on some other planet, for destruction!

Laszlo: We need to find our home and survive in our home, and find the harmony the cosmos is expressing in its various ways.

Tobias: Indeed.

Laszlo: If we can find it, and contribute to it, we act on the noblest aspiration available for our species.

Tobias: I agree with that. And that brings us back full circle to your home right here in Tuscany. If those same constraints that you suggested earlier with respect to the implausibility of the "randomness" argument exist—given the 13.8 billion light years that the universe has had to come up with this idea, this reality called life—and if our noblest goal is providing for the neighborhood (that surrounding village, city, or countryside we all inhabit) with love; and if we can somehow contribute to the great good here on Earth harmoniously, then we will have fulfilled, as Aristotle put it, the *eudaemonia*, the ultimate entelechy, peace, or end result of human life, the achievement of true happiness. A kind of Utopia. Probably "the Utopia of a Tired Man," as the writer Jorge Luis Borges put it; of very tired people. And ironically, the same aspirations have been jaded by a serious cynicism at large in a world rankled by ongoing violence in so many sectors. We've lost count.

We've seen it, for example, in the case of Sir Thomas Moore. Norway, a country of four million people, just waiting to implode on many levels, not least of which is the killing of whales and harvesting oil from the oceans, was, in a sense, his model for utopia. That was Moore. Samuel Butler had his own Erewhon, or "nowhere"— that is to say, an imagined utopia—located physically near the city of Christchurch at the Mesopotamia Sheep Station on New Zealand's mountainous South Island in the 1860s. He lived there for four years before coming back to England. His utopia had collapsed, just like the "collapse" dissected in Jared Diamond's book of the same title, where he analyzes places like Easter Island (Rapa Nui) and the state of Montana with its extensive ground water and river pollution. We've seen that with every attempt to achieve harmony there is a grim shadow that appears. This is one way of telling the human experience. Authoritarian power, fascism, Nazism, and now a new kind of rise to power in the political landscape of the United States. I'm not suggesting that there aren't other, more sanguine stories to tell.

LASZLO: Let me just interject one point about "attempting harmony." If we access the deep harmony of the cosmos, the harmony that is in every cell of our body, we do not make mistakes. We seek to become one with the things embraced in that harmony. But if we are seeking instructions, recipes, we are always in danger of coming up with the wrong information.

TOBIAS: Certainly. Gandhi, when he gave that famous speech to the wool industry leaders in Manchester, said, look, we need to return to spinning, we need to produce our own wool and our own cotton. That was a form of instruction. If you look at the world up until recently, it was "a walked world," as George Steiner once described it. Dante walked his world, Shakespeare walked it and, yes, rode a horse through it occasionally, but basically our species has walked the planet until just recently. So the rules governing our behavior have been largely dominated by the existing technologies, which in the case of our ancestors up until most recently were comprised largely of our own bodies. That's what we've had to deal with. Now we have surrogates for our own bodies, in a sense. We have seen the hybridizing of crops, iron and copper mining, the draining of swamps, the fast-freezing technology to preserve fish, animals, and vegetables we consume, and the transcontinental shipping in freezers. We have come up with air conditioning, we can tap something on the wall and we get light, we don't have to build campfires, and we don't have to find fire.

My obvious point is that we have blueprints, we have instructions. Is it any wonder that how-to books tend to be the best sellers of all time? We seem to be a species, in my opinion, that is seeking help and desperate for answers. A child looks to its mother for help and affirmation and expects it, takes it for granted, and knows quite angrily and sullenly when that help is not there. If that child is orphaned or if that child does not get water, food, or is abused, she or he becomes a statistic. We know the dysfunction only too well. So, a child looks up at his father with an inquisitive glance and says, "Daddy what should I do, teach me what to do. What am I supposed to do?" So, as a species, who do we look to, what are we supposed to do?

LASZLO: There is no need for seeking help from the outside, that's the point I'm trying to make. The great prophets, the great spiritual leaders have always told us we have to look inside because we are an expression of the process in which we are participating.

TOBIAS: That's well put.

LASZLO: Seeking something, a set of instructions to manipulate others, is a dangerous game to play. We have something in us that is expressing itself in the widest reaches of the cosmos. It is the basis of health, balance, and harmony in the world, and it's also instinctive, so trust your intuition. My sense is that this intuition is more important than reasoning. You have to use reasoning because it's not enough just to have intuition, but without intuition your reasoning is nine times out of ten likely to take you to mistaken ways. I give you a well-known example that concerns the Rubik's Cube. Fred Hoyle, whom I mentioned earlier, said that if you make a random trial move once per second to try to organize a randomly scrambled Rubik's Cube to get it into perfect shape, you will get an x number, which is expressed as a power of ten. If you translate that number into years, we get twenty-six times the age of the universe.

TOBIAS: That's the case for the blind person, but a normal person could get feedback . . .

LASZLO: With feedback, whether the moves are right or wrong, the Cube can be unscrambled in about two minutes. This shows that you need to know whether each move is right or wrong. Randomness is simply not an option. It doesn't explain what we find in the world. There is nothing in the world that woud be entirely a play of chance.

TOBIAS: I'd like to come back to what I referenced earlier and have, in some ways, been circling. The polarities that have hounded our species can be easily dismissed simply by waking up on that side of one's perch with a disposition to be happy, to try to be virtuous, and to live the life of philosophers and ethical thinkers and people of faith. So many have strived for that paradigm throughout recorded human history. There's no question that the capacity among our kind to be tender, generous, and unstinting is there. And that, in my opinion, reflects a bigger picture in which there is gentle cooperation and predominantly herbivorous behavior throughout the biosphere. Indeed, many years ago in my research throughout parts of Kenya during which I amassed considerable empirical and primary source data, I was able to estimate with some level of clarity that for large vertebrates, not birds or invertebrates, the ratio of carnivores to herbivores is something like one to four thousand. In other words, for every meat eater, there are some four thousand vegetarians. So from that I make out a rough but poignant extrapolation for the entire world. It suggests that for large vertebrates like ourselves, we do have a capacity to live life sustainably without inflicting unnecessary cruelty on others. To live and let live. And what surprises me is that this contagious ideal is not more infective, does not catch on with far greater frequency.

It amazes me that we continue to fight like idiots and kill like the monsters that the world's audiences thrill to. And I'm confronted by the inevitable conclusion at this stage in my life that unless we change this, we're doomed. On many days I wake up and feel that way. It's one of the reasons my wife, Jane, and I chose not to have children. We didn't want to contribute to the population boom—and with each potential child, more consumption, more human violence. So despite all my assiduous efforts in helping individuals, for every individual I have some humble part in saving, there will be fifty others, fifty billion others, that will not be rescued or spared ruination at the hands of *Homo sapiens*. Fifty billion others will be slaughtered.

And so of course we have nihilistic fatalism such that I want to light a match and not scream out at the darkness. But mostly the temptation to shut the door and listen to night's symphony . . . it is a temptation. Because I have the luxury to do that. Of course, when I try to actually fall asleep, I have serious insomnia, which I have had most of my life, since the age of three, when I first confronted a caged wolf. Each and every night I have trouble sleeping because I'm thinking of all those who don't have the luxury to do what they desire, for whom sheer survival is a ghastly challenge. I cannot escape the demons inside me. And that is the dialectic. It is, for me, certainly a chronic problem. I see every night the faces of all those wolves, all those cows, and pigs and turkeys and chickens and endangered species, extinct species, individuals, the ones I could not help.

LASZLO: If you extrapolate from the situation today and do it consistently, you get to a dead end. there's no question about this. The meaning of sustainability is that you cannot keep doing the same thing you are doing without the system collapsing. But this is not a linear process. What I've come across in my discussions with the Nobel biophysicist Ilya Prigogine and many others is the idea of a bifurcation; it's there in nature, and it applies to every complex system. The system is stable—or semi-stable—for years, or thousands of years, or entire epochs. Then its reaches a point where the fluctuations, the tensions in the system, rise to a critical point, and then the system cannot continue to evolve on a stable trajectory. It can collapse and disaggregate to its stable components, but it can also reconfigure itself. The outcome depends on the kind of information that codes the system. If that information is genetic, you can't change it. But that's not the problem because even genetic information is not what finally codes the behavior of the system. The key factor is the epigenetic system, the system that switches genes on and off. That system can "learn," and its learning is to some extent transmittable. However, if it's rigid, at the bifurcation point the system collapses and disappears. There are systems that can avoid that fate.

The majority of the systems can be reprogrammed or can reprogram themselves. There is hope. We are approaching a critical point, but a lot of things remain possible. Except, of course, remaining the way we are.

TOBIAS: Indeed.

LASZLO: We can ask ourselves, are we heading towards a collapse, a breakdown, or are we heading towards a reprogramming, a breakthrough? Right now, with 7.4 billion people on the planet growing so fast, and the weight of the load we are putting on the planet, our life-support systems, are totally unsustainable. We're using our sophisticated technologies just to maintain ourselves at the edge of its stability, just so our system doesn't collapse immediately. These technologies are now coming to the limits of their effectiveness. That's why we're having all the species extinctions and massive outbreaks of violence, ecosystem collapses, and the threat of a global planetary population collapse. As it is now, the system is not sustainable. If you concentrate on the system as it now operates, no wonder you have sleepless nights. But the other side of the coin is that our systems have incredible resilience and the capability to restructure themselves from the ground up. Now it may be beyond a given point, beyond the so-called critical mass. This calls for a great deal of sacrifice.

The critical mass is relative to the level of stress in the system and the level of its sustainability or unsustainability. The greater the stress, the greater the instability, the smaller is the critical mass. Because then, any small push can flip the system over, whether to a breakdown or to a breakthrough. If a system still has resilience, still has residual sustainability, then we need a very large critical mass. The fact is that the human system is now becoming so incoherent, so prone to collapse, that relatively small groups can change its dynamics. Margaret Mead said never doubt that a small group of people can change the world. I would add that small groups of people can change the system when the system is critically unstable. Now we are moving toward that point.

TOBIAS: What evidence do you have for the bifurcation process?

LASZLO: All the great changes in the twentieth century were bifurcations. They were triggered by relatively small groups of people. The most visible instance has been Hitler and Lenin in the twentieth century. Both led peripheral groups. They invaded the system with ideas that were irrelevant and unrealistic at their time. But because the system was so unstable and there were no alternatives, and no competing movements sufficiently arrogant to be competitive, they could take control of the dynamics of the system.

TOBIAS: Two examples jump out at me in terms of a paradigm shift, neither of which were very pleasant. When Cortes entered Montezuma's capital of the Aztecs in the Valley of Mexico City in the 1520s with a very small group of soldiers, he wiped out the Aztec empire. They not only vanquished them physically and morally, but to add insult to injury, Cortes' troops also burnt Montezuma's four-story aviary. Symbolically and in reality, the Spaniards burnt thousands of animals alive, including representative bird species from throughout the neotropics—among them the resplendent quetzal. A horrible kind of scorched earth policy remnant was inflicted by a very small group of men. Another example would be the case of Rapa Nui civilization, Easter Island, in the biodiversity hot spot of the Pacific where someone, one person, cut down the last tree—and knowingly, at that, because you can see from the highest point on the island the entire 160 odd square kilometers and every tree on it. So you didn't have to imagine it, or calculate it. One could see (I've been to the spot) that you were absolutely undermining the survivability of your culture way out there on one of the most isolated places in the world. There would be no hope for children without trees.

So human beings can have these negative shifts, and when you cut down the last tree, that's it; when a species goes extinct, that's it. We've seen huge collapses of various food webs. We're seeing it now in the oceans—all twenty-two major fisheries—and we're way beyond the point of no return. I would wager that the majority of marine biologists would agree that we may well be beyond the point of no return for the oceans and fresh water systems, which means a lot of things to a lot of human beings, but it means much more to many other individuals of other species across the planet.

There is a widespread consensus amongst ecologists worldwide that the sixth extinction spasm, which the planet is now enduring, has placed in great doubt the survivability for much longer of the majority of large vertebrates. Indeed, one of the great minds in this deeply disturbing field (the tracking of biological extinctions), Professor Gerardo Ceballos of the Universidad Nacional Autónoma de México, has called the trend "biological annihilation," citing the fact that in just a little more than the past century some 50 percent of 177 mammalian species analyzed had suffered the destruction of some 80 percent of their population distributions. This information has filtered down from the ecosciences and prestigious publications like the *Proceedings of the National Academy of Science* (where Ceballos first published his conclusions) to the popular media, in this case specifically to *Wired* magazine, and an excellent overview by journalist Alexandra Simon-Lewis. (See "Earth Has Entered into a Sixth Mass Extinction Event," *Wired,* July 11, 2017.)

The heat island effect is escalating beyond anything we anticipated. Some of the calculations, with respect to rising sea levels, are now suggesting that in the worst-case scenario, a scenario which is now on the table, ocean rise could translate into seventy-five feet, not the three to nine feet originally considered. These are staggering prospects that could result in the loss of Antarctic ice shelves, of Greenland (the latter breaking a record in 2016 for warmest temperature ever documented), the extinction of large vertebrates in the northern and southern polar regions, complex changes to ocean salinity that will inevitably result in major changes to weather patterns, and on and on. So if, in fact, we have already gone beyond the point of no return, then the question perhaps is one of triage, which is not a word conservationists ever want to mention or deploy, because it's nearly hopeless when we're at that point. It would be like E. E. Cummings as an ambulance driver on the killings fields during World War One—recognizing that you have to leave some to die, a kind of *Sophie's Choice*, essentially.

So, short of that, if we want to posit hope, a positive disposition, I don't see how the knowledge that there's a system inherent to the Big Bang that has touched every particle in the universe will help us in our current predicament. The knowledge is fascinating; it encourages me to want to believe that there is a force greater than our own, but do I know how to enlist it in my everyday life to make things better? Maybe it makes me feel better for the few minutes I'm reading about it, which is a most ephemeral and personal luxury, but I'm still going to go to bed and suffer a sleepless night. Even though I know that change could come very fast.

LASZLO: But that's precisely what I want to say. Change can come very clear, very fast. It can come in the span of half a generation. It can come in the form of young people intuiting and internalizing the kind of values and life-principles that can lead to a sustainable world. That is the only change that makes a lasting difference.

Tobias: Look at the Club of Budapest's meeting of December 2014. There was enthusiasm from a very interdisciplinary mixed assemblage, including individuals who embraced and celebrated the social media revolution and the unconditional love revolution that is not just a phrase but a movement that's manifested (by such individuals as Mata Amritanandamayi Devi, or Amma as she's commonly known, and in the noble tradition of Mahavira, Saint Francis, Gandhi, and many others).

I joyously admit that just within the United States there are over twenty thousand animal rights organizations. This would have been unthinkable when Henry Spira first came up with the anti-vivisection arguments in the late nineteenth century or when in the early 1800s Lord Thomas Erskine forcibly took some of his colleagues in Parliament by the collar to wake them up, leading them outside and pointing his finger at the abuse being meted out to horses just outside the gates of Parliament. They were shocked, as if they had never seen it before. Because Erskine was able to break through to them, one of the first animal protection pieces of legislation in England was passed. He shook them out of their stupor, their confirmation biases and complacency that they had adhered to in terms of simply not ever really seeing before, despite having probably seen such abused animals every day of their lives on the streets of London.

Sometimes people need someone to wake them up with a "Hey, wait a moment" minute. For instance, "Don't you know there's a stop sign there?" Or "Didn't you notice that your wife is crying?" "Didn't you see that your children are hungry?" I recognize that people are vulnerable to change, in the best sense. And that when change happens it can be permanent.

Laszlo: It can become permanent very fast.

TOBIAS: Very fast. Equally fast, we know that the weather can change. And the anomalies are now the new norm and most people realize this.

LASZLO: We are in a race.

TOBIAS: We are. It's James Dean in that game of chicken in *Rebel Without a Cause.* Do we stop before the cliff? That's the question. We have the ability to put on the breaks, we just have to do it. To be or not to be comes down to "to behave or not to behave appropriately to the needs of the time."

LASZLO: We have to find the way forward.

TOBIAS: Yes. And we have to have the humility to turn around. And to embrace the wisdom of our ancestors in many regards, and to recognize the disappearance of indigenous peoples, languages, and perceptions . . .

LASZLO: This is why, if you were to ask me what is my aspiration, my contribution, I would say to help people to recognize the wisdom of our ancestors notwithstanding the now obsolete ways in which they have expressed it. Wisdom has to be credible, and credibility today comes from science. It comes from updating our thinking to what I call the new paradigm. If that wisdom can reappear in the framework of reasoned deductions from a wide range of observations, people can say, "Ah-ha!" They recognize it. The wisdom of our ancestors was true wisdom, but we don't need it in the symbolic terms in which they were expressed at the time. We can re-express them in scientific terms, and that would give it that extra credibility that would make all the difference.

TOBIAS: Yes, no question.

LASZLO: That is the way to make change. It is to realize that we have valid instincts, and they are not just imagination but deeply encoded in us and in the universe. If we reach this recognition, we could have a chain-reaction among the young people whose world this will be. They would recognize that this is a different world and we are different beings from what we had thought. We all have to revise our idea of what is "mind," and what is "matter" in the universe and what is the interaction between them.

TOBIAS: The Dalai Lama has told me that it is his belief that the redemption of the Tibetan peoples will not happen until this generation of leadership in Beijing dies out. Conversely, on the upside, you have not just the younger generation, the young Silicon Valley billionaires who are actually committing 50 percent of their wealth to helping the world, but you have older generations like Warren Buffett's son, Howard, who has invested in the ancestral practical wisdom throughout Africa of raising crops in a sustainable manner. This is in opposition to the fact that in the United States an astonishingly high percentage pf people believe that we coexisted with dinosaurs, and that climate change doesn't exist.

You mentioned the importance of revivifying science. The reality in the United States, which I'm using as a bellwether for the planet in some regards, is the fact that STEM (Science, Technology, Engineering, and Mathematics) receives some of the lowest-levels of funding. The Obama Administration tried hard to encourage and provide federal funds for the teaching of good science, but it's a struggle. As for his successor, he appears to have zero interest in, or understanding of science.

I remember that when Rita R. Colwell was head of the National Science Foundation around the time of the Millennium Conference (that UNESCO, under Federico Mayor Zaragoza, was involved with in Budapest in the year 2000), one of her biggest laments seemed to be the scientific and ecological illiteracy at large and the insufficiency of funding for basic science even in the NIH (National Institutes for Health) for holistic medical research in oncology and in other medical realms. The population at large gives very short shrift to science.

LASZLO: What you're saying is what we have been saying all along, that extrapolating from the current situation brings us to a dead end. The bifurcation is coming precisely because the current processes lead to a dead end. Hopefully, this will become evident before the crisis actually hits.

TOBIAS: We are at the twelfth hour.

LASZLO: We are certainly very close to it.

TOBIAS: We see this with eutrophic structures. A lake on the so-called twenty-ninth day when it is starved of oxygen is going to end up with an impoverished ecosystem. Many have projected what the earth will look like, even in what year certain species are calculated to go extinct. All of the expanding dead zones throughout the world's oceans, desertification, deforestation—these are all woes that the public now yawns at by way of a kneejerk reaction because they are so bored with hearing about them. They have grown up listening to how many areas equal to football fields are destroyed in Amazonia every minute. It has become a cliché, it's glib, it's something that we now live with every day in our headlines (now it's one tropical rainforest the size of Belgium being obliterated every year).

We learned as children in grade school that grasshoppers habituate to their own deaths, if not to extinction. They don't see it coming. We know that the frog becoming habituated to boiling water is a myth. That never happened. But in some symbolic sense we're inured to the crises. It is often said that it's going to take this or that disaster to elicit a response that is positive, and we're seeing it in micro-junctures of depravity in South Sudan and in Haiti after the earthquake. Where and when the world responds to tears became the capital of NGOs. We've seen it in the case of Japan after any number of disasters—Fukushima, typhoons—how the Japanese have always come together to help each other with outstanding cultural generosity and spirit and humbleness. And you intimated it in the welcoming culture of Italy.

I have a positive disposition or I wouldn't continue endeavoring to do everything I can to help mend the world, because it's not pleasant most of the time to be on your hands and knees seeing things that are desiccated or dying. Nobody likes to see a beloved tree they've planted collapse because of lack of water, or from lack of nutrients or destroyed habitat, or any number of negligent or impactful human behaviors. The bottom line is that nobody wants to see suffering or to suffer. This is a universal principle that every great spiritual or ethnic or artistic culture has embraced, and it's a starting point for me. It's that communion that is inherent to ourselves. I know of no organism in the world that wants to suffer. Buddha said that life is suffering, and it's a very obvious equation: If we are to redeem our species in time we need the input of as many as possible. We need, I believe, to provide points of access for as many people as possible. We need to engage as many people as possible in a conversation that is practical, but also brings tears of joy.

LASZLO: The most practical input we can make is to spread the recognition of the dead-end toward which we are heading, and of the possibility of finding another way.

TOBIAS: I totally concur.

LASZLO: And then the change can come of its own. If the species is viable, then it will come. It's not sure when it will come, but what is sure is that it is going to take a traumatic transition. How long this transition will take is unclear, but it may entail a drastic reduction of the population. We are talking about billions of people. We don't know how to envisage it in realistic terms.

TOBIAS: You've said it well. It is logical that we should experience the pains of childbirth to bring forth a new paradigm, the labor of a bifurcation. I have personally read the runes, so to speak, in divining the many biological thresholds, constraints, and far perimeters and I think we're there. Everything my training, but more importantly my feelings, has taught me is that we're going through it now, this radical paradigm shift, a chilling but informing bifurcation.

DAY   TWO

Afternoon, on the Front Patio

TOBIAS: A thought occurred to me, reflecting on what we said this morning. This is no longer about putting off the obvious and embracing the logic of disasters occurring simultaneously across the earth, much like the millions of lightning strikes happening continually. It is not a question of tomorrow. We are there now. It is happening, "it" referring to a bifurcation whose all poignant centerpiece comprises the astonishing and horrific death of the world. Some see that as a "whimper," others as a "bang," to quote T. S. Eliot.

But make no mistake: we are in the throes of the Anthropocene. We have some obvious legacies. We know that the last thylacine, popularly misnamed the Tasmanian tiger, went extinct in Australia in the mid-1930s. We know from an array of imagery what the world's last passenger pigeon, Martha, looked like prior to her death at the Cincinnati Zoo in 1914. At one time the most multitudinous bird ever to grace the skies of North America, billions of passenger pigeons could once be seen flying to their nests across Wisconsin at sunset. She was a beautiful pigeon, truly a beautiful creature. And the ardent ornithologist John James Audubon in his late career assumed this was one bird that could never go extinct. We now understand only too well that these tragedies can and do happen; many—most scientists and educators—understand that this disaster is waiting to happen, and perhaps is already happening.

On top of that, it is only a matter of time before another dirty or smart bomb (certainly not smart) will get into the hands of a terrorist group, lone misanthrope, or infantile dictator. There will be another catastrophe. The Carter Center in Atlanta has documented more civil wars simultaneously erupting like wildfires than at any other time in recorded history. So, we're going through it, all right. We're possibly, in ways yet to be perceived or fully understood, expressing the anguish and personal nihilisms associated with the Anthropocene, that vast extinction of biodiversity. If the bifurcation we have spoken of is up to it, I rather insistently pray there is a new beginning and it is a good one.

LASZLO: When there is a choice, you have the possibility of a bifurcation. Bifurcation entails, and is based on, choice. As I always say, evolution does not come with a guarantee of success. It can lead to extinction, and in the majority of cases it has led to extinction. The plus sign on our side is that we are able to perceive it, and if we are able to perceive it we can accelerate the process. This could lead to our salvation. This is not in changing the directives but in changing ourselves. That is much more reliable than trying to figure out the strategies in terms of inputs and outputs and consequences. That will come, but one first has to feel the direction in which one wants to move. First, we have to become a different human being.

TOBIAS: I've said for decades that evolution does not condemn or liberate us. Only our choices can do that. And most recently, my wife, Jane, and I have written a book entitled *The Theoretical Individual* that examines the philosophical basis and evolutionary contextualization of choices in humans and in other species. It discusses the hybrid futures we can expect and the tired notion of Utopia that, in ecological terms, is the perfectly balanced line between the X and Y coordinates. In other words, nature herself, always in flux, knows well the basis for change for inputs and outputs. She (Gaia) has explored these central pillars of ecodynamics, of all biology, including molecular biology and macromolecules (read: units of information) of course, for billions of years. Our species, without acknowledging or even thoroughly understanding it, has been walking on a red carpet since our inception as hominids. A zoological palate of extremes, compromises, experiments, homeostasis, concord, discord, aspiration, competition, bliss has been feverishly exploring every niche and cranny of creation prior to our rather abrupt, if not outright rude, arrival.

When we speak of first becoming human beings, I would add that there are countless noumenal qualia with which our bodies and psyches must grapple. There is the heavy burden of Darwin's thinking shaping, in part, myriad forces of primate evolution that we as a species have subsumed. These are multiples of currents, views, anatomical realities that are with us. We have an internal compass, a species-wide biological clock that senses that direction, or those directions. Within that sensate itinerary, our place in evolution, we either awaken to the miracles of life, of remarkable biodiversity, or we risk going extinct. This is not a solo journey, but an ecological journey here on Earth. It can't be overemphasized: All life forms are interdependent.

LASZLO: A human being does not have a preset, unchangeable nature. We are part of the web of life that evolves, and we ourselves evolve. This evolution has now gone off on a tangent, but it is a tangent on the surface only, because the majority of the people on this planet are not pushing in that direction. They express an enduring element of humanness. They want to, and can, live together. They can live with nature and the planet. But they are placed in situations where this is not the preferred way. This is destroying nature, life on the planet.

TOBIAS: It is shown by the poverty that is so endemic amongst billions of humans, as Amartya Sen has so persuasively described.

LASZLO: Poverty and deprivation, and the ignorance in which they are rooted.

TOBIAS: Well, ignorance yes. But on an empty stomach it is very difficult to read Plato or Martin Buber.

LASZLO: On an empty stomach you could still be moving with the intuition that motivated Plato, that there is a higher reality than the one in front of our eyes, the intuition of oneness with nature and each other. That intuition is the key to our collective survival.

TOBIAS: There is the true story of the man in India whose wife died because she could not be taken to a hospital quickly enough in an emergency situation because there was no road through the mountain near their town. So for over a decade he set about, without a dime of government funding, hewing with simple tools the very three-kilometer road that would have saved his wife. And he did it, this bereaved husband who was so committed to preventing this absence of infrastructure from ever again endangering the life of a loved one. People can do amazing things with a shovel and a pickax. There is a film about him now. Some people have what it takes. There is another individual in India's northeast state of Arunachal Pradesh who has spent most of his life replanting the denuded, abandoned sandspit with native trees. Large, charismatic megafauna have returned—tigers, elephants, rhinos, etc.—and it's all because of just one middle-aged hero who thinks nothing of planting routinely, day by day.

LASZLO: It's the vision. You know what you want to do because you feel that it is something that has to be done. And when all is said and done, this vision is the best hope for humankind.

TOBIAS: I can't believe that our species won't get its act together. Mammals in particular, and vertebrates in general, have embraced the divinity of community. Either in spite of, or because of evolution. It's a big question. We know, however, that the conscience—what the ancient Greeks knew as synderesis—segues seamlessly into that domain of the love of or even reverence for nature. The Greeks called that passion physiolatry, the love of nature. I believe it is instinctive throughout the biosphere and even comports with contemporary views with respect to biophilia, a shared passion for life upheld by all the diverse actors on this global stage.

I said earlier that no being wants to suffer. This is a universal principle that transcends physical characteristics. The earth has given rise to, and her life-affirming principles are commensurate with, both the largest known beings—mile-long carpets of moss, aspen groves, blue whales, elephants, and so on—as well as submicron-sized cellular displays and behavior. Most organisms that we have thus far encountered share with humans much if not all of the lexicon of behavior, at some level, whether we get full or meager glimpses. We see throughout the tree of life beings that grieve over their dead, form strong social bonds, display some of the most contagious affection for one another we can imagine, and have extraordinary languages. Some may wield more verb tenses in a single neotropical canopy (I am referring to macaws, for example) than we can possibly imagine. Birds dream (we see their flickering eyes); and they have every right to benefit from their millions, tens of millions, hundreds of millions of years of brilliant trial and error and success and failure, but ultimately success, of learned experience.

In attempting to express their worlds by way of human language and concept, analogies are awkward and often ill-fitting. But we sense that they—the worlds of birds, marine mammals, social insects, vast forest habitats, coral reefs—have created a sustainable world for themselves that is dominated by love and tenderness. And in some cases, by absolute forgiveness. Of course while people have observed what they believe to be these qualities, we're not sure what we're really observing. But because we relate by way of accustomed metaphors with which we are comfortable and at home, I suspect these are accurate depictions of a behavioral biosphere. And in my view it is absolutely critical that we learn from this multiplicity of zoological gazes and epiphanies.

And I can't believe that with such exemplars, with such gurus all around us, we who presume to know a thing or two, surround ourselves with books, sit and discuss great ideas, cannot get it right. I have to believe that we can.

LASZLO: That's the coevolution of consciousness. What is emerging among these species, the higher more evolved social-ecological species, is the recognition of their oneness, their nonlocality, or, to use another term from science, their "entanglement" with each other and with the world around them. The intuition of the direction in which evolution is unfolding. We humans need to recognize that our own evolution has been on a diversionary path for the past several hundred years, and perhaps longer, but certainly at least since the beginnings of the Industrial Revolution, which was based on materialist, reductionist, fragmentary thinking. This is what has to be corrected. I am convinced that what will really help is a paradigm shift, not just in regard to our intellectual thinking, but in our being, our whole psychological makeup. Above all, in regard to our relations to each other.

TOBIAS: I'd like us to reflect upon what those ideals, that idealistic thinking is. I mean, I look at the Amazon and I see the ideal. I look at the Cecina River Valley right beyond your and Carita's house right here in Tuscany and I see it, right outside your windows. And I point out frequently, because it points itself to me, that in our very backyards—whether we're speaking of Manhattan, rural Indonesia, or the Mongolian wilderness—there is an idea of nature that has become the ideal of itself. The landscapes are crucial to expressing this universally felt ideal. An Impressionist painting of a picnic places us there. Or that great image from Kurosawa's film *Ran* comes to mind, of the old tormented man on the cliff, much like Shakespeare's King Lear.

In all such metaphoric instances I can still find a microcosm of perfection, of great joy, beauty, and solace that nature induces. All those superlatives that were the sublime underpinnings of such poets as Rilke, Wordsworth, and Shelley. Truly, every shade tree gives me great enthrallment. Right outside here today, on a Tuscan hill, it was quite hot, but by simply moving three feet into the shade of a tree the relief to my person was profound. This is a simple instance of the critical role the environment plays in maintaining the homeostasis that we, as biological beings, often take for granted. We know from recent studies, for example, that the human metabolism is directly correlated with heat and with cold, with a humidity index that appears to be fixed in terms of fundamental comfort zones at around 84 degrees Fahrenheit. After 84 degrees, our bodies start to compensate—frantically, I might add. We sweat and metabolic activities begin autonomically with a rapidity that lets us know we are somewhat uncomfortable, then that we are dangerously so. After 104 degrees, we're near death.

In other words, there are huge GDP and psychic performance variables conditional upon climate change that will utterly alter all civilizations and the economic language. A tree is everything; its shade a miracle. Without it we are dead. And we have killed 50 percent of all forests on the planet. By most accounts, in 2100 climate change will have exposed some 75 percent of the human species to horrible threats, and that is without even accounting for the atrocious feedback from other declining species and their habitats. The ecological downward spiral of multiple trophic levels cannot be calculated, but we can ascertain enough of the puzzle to recognize utter disaster across the planet. And it is happening far faster than all the Nobel Prize-winning scientists ever anticipated. It's horrifying. Every species has its own metabolic liberties and constraints, but the totality of the biosphere is conditioned by over four billion years of evolutionary lessons that we ignore or overstep at our extreme peril. We know this or we should know this. The great sadness is the fact of our placing that peril in the face of so many others.

Such examples should teach us to plant trees—as many as we can and as fast as possible. The more than one million NGOs around the world are trying to plant trees, the seed of an idea, to give a child hope, whatever the cause, the mission . . . there are a lot of people out there doing a lot of good. I have to believe that the ecological restoration of the earth, the great redemption, is at work in human hands. Everyone else (referring to every other species) has got it right. But we have got it wrong. So unless we quickly become agents of resurrection within the context of our one species having inordinately inflicted harm upon this precious planet we cohabit with all those other one hundred million species, times millions of individuals per species, we are paleo-history. One more biological mistake among millions of them. But if there is this collapse, which in my mind is unambiguous, I think at this point we should think of ourselves as on a spacecraft that is going to crash land. The questions cascade: Are our seatbelts tightened? Have we given any thought to what we're going to need once we crash, how we are going to reorganize ourselves, and when we do rebuild, how we can do it better? Distributive justice. Is it comprehensive to *Homo sapiens*? How shall we reconsider our obsolete notions of rights and duties in a biosphere so obviously interdependent? These challenges are thrilling, desperate, and of an urgent priority if we are to survive ethically. That is quite a different proposition than simply surviving, I might add.

LASZLO: I would use this metaphor, but change it slightly. We are on a spacecraft that we need to restructure in such a way that it has enduring sources of energy and enduring resources and a sufficiently balanced system of distribution—so it can be maintained in working order, and the crew can not only just survive, but flourish.

TOBIAS: That's optimistic. Especially in light of the biospheric collapse I've been touching upon.

LASZLO: Not moving blindly forward, but restructuring our way ahead: that's the task that awaits us.

TOBIAS: I want to make sure we're not simply on the Titanic rearranging the deck chairs, as they say. That to me is the ultimate nightmare.

LASZLO: Well, that's why I say we need restructuring in a quite fundamental way. Reorganizing our sources of energy and of our organic and inorganic resources and balancing our ecology. That is not just arranging our deck chairs. Doing these things are fundamental.

TOBIAS: Right.

LASZLO: It's not a question only of finding and giving the right instructions, or of knowing what you have to do: It's also about *wanting* to do it. Because actually we know a good deal about what there is to do. We know how to use renewable energies. We know how to redistribute basic resources. We also know, as you said yourself, that it's very hard to change your thinking on an empty stomach. We know that we have to have a minimum level of well-being on this earth to have sufficient momentum for the change that's needed for the whole system to survive. We know all that. What we are lacking is not detailed knowledge but the realization that we can change our ways, not just because we need to, but because we can't live otherwise than by changing them. That's our survival instinct.

TOBIAS: It's an old cliché, ultimately, that we all heard as children: you can lead a horse to water . . .

LASZLO: Yes, but what we call drinking is an inner and innate knowledge. It is based on the evolution of our consciousness. Based on the consciousness that is emerging in humanity and in the universe. It is a consciousness of connection, of coherence, of alignment expressed as oneness and love. It's the kind of consciousness that whales have developed. It is strong enough to be expressed in behavior even if it is not rationally recognized. Now, we have the possibility to recognize it and to use our recognition to spur a change in our behavior.

TOBIAS: That's a question mark for me. I think one of the most powerful experiences in my life was doing a feature film on Amma, the hugging guru in Kerala. What people most admire about her is not just that she hugs, but her seemingly inexhaustible capacity to keep on doing so, and hugging everyone, no matter their station in life.

LASZLO: That is right.

TOBIAS: I'm hoping that the consciousness of connection, of coherence, about which you speak is indeed the ultimate catalyst emergent in our souls and psyche. That the collective unconscious, the atavistic residuum of all that has ever been and will be (as Carl Jung and James Joyce understood the challenge), comes alive on the focal point of our own survival and the pertinacity of all those with whom we cohabit the planet, whether blue whales or cockroaches. Every member of the zoological family called life. That seems to be the central pillar of Amma's inexhaustibility, from what I have glimpsed up close and personal with her at her ashram in Kerala.

THE VENUE FOR THE TUSCANY DIALOGUES

+ + +
+ + +

The Villa Franatoni and the Historic Valley of Cecina

The Villa Franatoni's front terrace and upper view-terrace

The study and the views from the garden at the Villa Franatoni

The fountain at the Village Franatoni (opposite page)

On the Cecina River below the Villa Franatoni (above)

The footpath down to the river

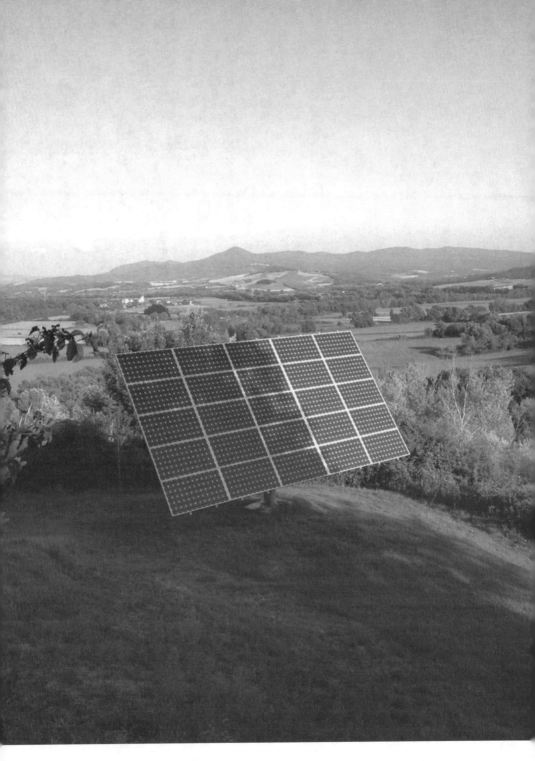

The solar panel that follows the sun and powers the Villa Franatoni

*A typical house of the Cecini Valley, with room
for people above and for their sheep and cows below*

DAY THREE

Morning, in the Study

TOBIAS: Something I came up with last night is what I call Synthesis Oppositional Syndrome. It is what the Romanian philosopher Emil Cioran emphatically aroused in me on reading his two seminal texts, *A Short History of Decay* (as translated by Richard Howard in 1975) and *The Temptation to Exist*. With every redaction of human evolution comes the inevitable counter-effect—the devolution. As an anthrozoologist and a human being I am bound by laws of my own personality and ontology. Those laws compel me to want desperately to envision, hope for, and then most importantly strive to engender the stable emotional, conceptual, and pragmatic conditions for transcending contradictions in the here and the now, but also in the hereafter. By analogy I am referring to atomic stability and chemical homeostasis. By that I refer to the intergenerational proceedings of life long after my own meaningless death, the realities of the very real tomorrow of my successors, the myriad of biological individuals to come, to whom such concerns are both emotionally and biologically vested. These convictions are as bound by evolution as by our own personal idealism. I'm speaking of our genetic future, the neotropics, the Amazonian lungs of the earth, the coral reefs, the millions of microbiomes and climes, populations, and trillions of individual sentient beings here in the semiosphere—one hundred million species all communicating with each other.

Last night as I made my way down towards the splendid Tuscan river valley by starlight, essentially courting that fine science of getting lost, I realized that this volition on my part, which has been the passion of my life, has always been shadowed by constraints, conflicts, every version of ethical challenge. I call such threats and concomitant opportunities Synthesis Oppositional Syndrome, or SOS, which is its ironic acronym with connotations of the Anthropocene, this human induced global problematique. I believe, and many others do as well, that we are confronted with something altogether new, given the unprecedented stakes of our time. In this framework of thought one must find the wherewithal beyond the veils of perception. According to one's own tolerances, attitudes, and unique personality vectors, this SOS invites us with an unstinting urgency to grasp, to get beyond, say, Oswald Spengler's analysis in *The Decline of the West*, or Nietzsche's demise of God.

Last night I took hold of the bark of a tree, probably half a mile beneath your marvelous, elusive, mysterious, and elegant villa, standing in mud on the river bank, completely in the wild. Though it's a new wild for me, all wilds have things in common that welcome me—birds and frogs and so many creatures I know. I was remembering that moment detailed by Jean-Paul Sartre's ultimate ecological existentialism in both *La Nausée*—the encounter of his most monumentally elusive Roquentin with the bark of a tree—and in his masterpiece, *Being and Nothingness*. I was, as the Mbuti Pygmies do, loving the tree, communicating with her. And thinking back to Sartre's distant cousin, Albert Schweitzer, who lay claim to a fundamental human pillar, namely sentimentality, which he implemented as a lifesaving doctor working on behalf of humans as well as wounded animals while living in Africa. As a sidebar, when I was a boy, I was invited to play Bach on the organ alongside Schweitzer in Lambaréné, Gabon. I foolishly declined to go. Had I been familiar with his fundamental philosophy, "Ehrfurcht vor dem Leben" (Reverence for Life), I would surely have taken the leap. I regret not going to this day.

But all of those thoughts touched me last night, a deeply personal, and I would add, revelatory, late night. It helped me to philosophically underscore this SOS concept, with its double entendre. A mere acronym, a phrase, yes, but it might well serve as a useful starting point for breaking through contradictions that we were touching upon yesterday. Namely, how to grasp this notion of a framework for stepping out of ourselves to embrace the other and thereby view the world really with a fresh lens. We spoke of revivification yesterday, of finding ways to achieve hope against this onrush of so much bad news—hence, this Synthesis Oppositional Syndrome, SOS. Transcending contradiction in the here and the now.

LASZLO: I don't think in terms of contradictions in nature. Contradiction is something that happens when you have a thesis and an antithesis and you seek a synthesis, the dialectical method. But in the real world there is nothing like an actual contradiction. There is opposition, contrasts, processes that go counter to one another, at least in the short term. In the long-term and global frame, I don't think there is anything that is opposing permanently and fundamentally the unfolding of the basic evolutionary trend. There is opposition and resistance in and around us, but that is part of a process that is strongly nonlinear. We are anthropomorphizing it by thinking that it is "seeking" the unfolding of the evolutionary trend present in all things. We have to find here and now how to resolve processes that go counter to the evolutionary process. It is a question of facilitating, promoting, hosting, and fostering a process that is unfolding, that in the anthropomorphized expression "seeks" to become reality. This is a task for a conscious entity that is able to perceive elements of the larger picture and place itself at the service of the unfolding that takes place in that larger framework.

Thus, as I said, I don't think in terms of contradiction, I think in terms of processes that can go counter to the harmony that is finding expression in the world. This is a Platonic idea, and it is akin also to Spinoza's thinking—that there is a search for the harmony, the synthesis that is pre-established and is seeking to emerge. It is our task to tender to and enhance that emergence.

TOBIAS: When you use the world "anthropomorphizing" are you suggesting that whatever we do as humans in our expressions, in our quests, we are duty bound to our species, not out of loyalty but out of the sheer presence in us and all around us of the inevitability—genetically, primordially speaking—of who we quintessentially are? That we are perhaps unable to break through those veils of perception and experience and to foster collaboration with that harmony?

LASZLO: Well if you recognize that those are veils—

TOBIAS: It's an expression . . .

LASZLO: I know that, but if you recognize that there is a veil or filter in the expression of evolution in us, we can make that veil thinner, reduce that filter, and enhance our vision. We can do this because the cosmic information is always reaching us. It is a holographic information where every small part contains the whole. When transmitted and transduced into electrical signals in the brain, it becomes available to us. Then we can recognize that it is filtered and reduced to allow for our survival, so we can maintain ourselves in the physically improbable state far from equilibrium—the state in which the irreversible processes can take place that enable living things to sustain themselves. This recognition is a survival imperative for human beings.

Beyond meeting the basic survival imperative, we can try to reach, as it were, for the stars. We have been doing that for thousands of years. We have been looking at the sky, looking beyond the horizon, asking what is there beyond our surroundings, and what does it all mean? That is an ongoing endeavor and leads towards an increasingly articulate comprehension of the processes that unfold around us, processes that can be articulated in different ways, symbolically as well as rationally, using one or another kind of symbolism and rationality, taking the Platonic, Aristotelian, or Baconian approach. Ultimately, we seek to grasp the nature of the reality of which we are a part and an expression.

TOBIAS: You used two words that caught my attention just now: *irreversible*, in terms of the life force, but also the *imperative* (overwhelmingly in terms of all life-forms) for survival. They seem contradictory, implicitly, because if life is irreversible then the imperative is diminished. We don't need to worry so much if the irreversibility is a fact. We can worry about our own personal survival, our family survival. And depending on the extent of our circles of compassion, we can worry about the plight of the remaining elephants, the 30,000 lions and 28,000 rhinos left in the wild, fewer than 146 kakapos in New Zealand, 50 Amur tigers, some 300 California condors. We also must admit to what is likely one of the most prominent causes of loss of life, namely, hunting. Very few rural neighborhoods anywhere in the world escape it. Italy herself is rife with the hunting of wildlife, both legal and illegal, including right around here, where local hunters seek wild boar and pheasants just up the Cecina River Valley. If it upsets us enough to act in favor of preserving life, that is a good thing. But if life herself is irreversible, the biological incarnation of the quanta originating in that Big Bang, then biological organisms are simply one phase of this ever-expanding logic and our efforts to save them from hunters and climate change and so forth might be inconsequential. But of course, human consciousness is part of that. Condor consciousness, from the slopes of Chimborazo in Ecuador to the cliffsides of Santa Barbara County, California, equally so.

You ask how consciousness arises out of a Big Bang, and it's a great question. But consider this: if life is indeed irreversible, which those who adhere to the Gaia hypothesis would certainly contend, then that same momentum must recognize, honor, and protect a remarkable story line. The earth has been experimenting by trial and effort (I don't say error), beginning with the anaerobic reality, and then ultimately permitting terrestrial life-forms to come from the sea and breathe oxygen, which didn't exist prior to that viable extent, leading to the big bang of biodiversity some 700 million years ago. After that came gymnosperms, "naked seeds"; angiosperms, or flowering plants; and so on. And then this irreversibility leading to an allegedly enlightened species that grasps it all. Well, if that is truly the story line there should be no fear. But in reality, we are honing in relentlessly on various doomsday scenarios. We've come all the way to mimicking nature with our herbicides, robots, GMOs, cloning, and private space ships aimed at Mars. Playing God, in other words. Talk about embracing irreversibility in ourselves. Inventing nuclear weapons. Tormenting subatomic particles and calling it physics. So our theoretical base refutes inhibition. We deny any collaborative sensitivity towards other life-forms, which I find to be utterly nihilistic, since it involves arming oneself with the inane pride of so-called human superiority over all our competitors in nature, whether bacterial or viral or other. All things negative—ego, greed, hatred, cruelty, torture—why should we care if the irreversible principle, theoretical or not, is capable of coming into being, given the many guises of that human history which has courted collapse? If life is irreversible, that alone should command such esteem, such loyalty, such a sense of privilege, that everything else is futile or Karmic. Why worry?

LASZLO: You are thinking here in the largest, cosmic dimension.

TOBIAS: Yes.

LASZLO: Evolution is bound to occur in the universe, no matter what we do. But there is a more elementary consideration that stems from thermodynamics, which is that every process that has to do with life has to be irreversible, because a reversible process means entropy, leading to a condition where nothing new can happen. In a reversible process from A to B and back from B to A nothing new has taken place. That leads to total randomness at maximum entropy. As soon as you add negative entropy into the process, you introduce an irreversible element. An example of such a process is lighting a candle. Once you've lit a candle it is going to burn unless you blow it out. As long as it is lit, it is an irreversible process. Now life can be seen as a multiplicity of candles; one candle lighting another, one process generating another. They interact and produce an irreversible, higher level process. Yet they are all obeying the second law of thermodynamics: they are all moving towards the highest level of entropy as individual processes. But together, in combination, they can be moving in the contrary direction. That they do is the miracle of life. That is the insight stemming from the work of the thermodynamicist Ilya Prigogine. It is the insight that in themselves entropy-oriented processes can give rise to an overall process that moves in the contrary direction: it maintains itself in a far from equilibrium, negentropic state, and can even grow and develop in that state.

DAY THREE

Afternoon, at the Upstairs Terrace

LASZLO: As we said last night, we need to remember the positive aspect of evolution. Its processes constantly balance the physical trend toward equilibrium. As we are beginning to appreciate that the universe is an inexhaustible system—we don't know whether it is infinite or not, but for all practical purposes it is an inexhaustible source of potential and virtual energy, and this energy fuels all the things that emerge and evolve in space and time. Whether it is rooted in dark energy or in dark matter, we know that it is there in the so-called quantum vacuum. That is, as I said, a quantum plenum, an inexhaustible energy source. If it wasn't that, the whole universe would be moving irreversibly toward entropy, it would be running down. But because there is this inexhaustible source, this fountainhead of energy that can be transformed into kinetic and chemical energy, we can draw and build on it. Irreversibility is a global process including many local processes that are running downwards. But globally, the universe is moving in the contrary direction—it is "running up," so to speak.

TOBIAS: The word "imperative" still troubles me.

LASZLO: We have a vast set of potentials at our disposal, and in our own best interest it is "imperative" that we seek to realize them. These potentials act as a bias built into processes that would be otherwise directionless and random. This is why the universe can keep going, because there is this tendency or bias toward structure and organization. It is the key to building on the "in-formation" that "forms" the universe. The information is there, but how we transfer it into the structures that emerge in space and time allows for alternatives. The given fact is that there are structures and they have options for continuing to maintain themselves, notwithstanding the irreversible degradation of energy.

TOBIAS: All right, then that conforms with respect to human behavior and our perceptions of ourselves, our daily lives, our conception of how it is we are going to survive this generation, bringing the theoretical down to the practical in terms of behavioral modus operandi. It entails a perspective driven by the regard in which we hold everything that we do. If we focus on life in its most rudimentary and simplistic highlights—we marry, we have children, we feed ourselves, we feed others, we go about our business, we hopefully don't wipe ourselves out as a species, we nurture life around us—it's a very simplistic, emblematic conditioning. But it's of brief duration, relatively speaking, according to the biological story line I earlier referenced. Indeed, a blink of an eye, but for us it's everything. And so it's the minute against this macroscopic illimitable. And as you've been describing, it is an irreversible performance of the cosmos that we're just beginning to grasp in some sense.

LASZLO: We can bring this down to the level of everyday reality. I am trying to use key words and key concepts to do that. What we know is that in the systems that are building up in space and time, the key condition of their emergence, maintenance and further evolution is the fine-tuning of every one of their elements to all their other elements, as well as the fine-tuning of the whole system created by this fine-tuning to the rest of the world. This means first of all the immediate environment, the sphere that exists in the framework of sequentially wider environments.

Ultimately every system in this world is what it is because it is finely tuned to every other system. The universe itself is a whole system created by the coherent relations of all its parts. This coherence means that every element is tuned to all other elements in such a way that what happens to one affects others in some way and engenders a response in all the others. So, what does this mean for us?

TOBIAS: Mind-boggling biological possibilities, for one, assuming we can somehow create an international, bilateral compassion accelerator willing to greet with honesty and respect all those sharing the world with us. That is a new and imperative law of physics, I would suggest.

LASZLO: It means seeking our coherence on multiple levels. Coherence of all the components of our own organism, which means the living cell and the cellular structures, the clusters of cells that make organs and organ systems, and ultimately the whole organism. That coherence is what we know as health; it is the coherence of every part of our bodies. Disease means that one part of the body is less coherent than other parts; in the extreme case, life is then no longer possible. In cancer one set of cells has become independent of the rest and has created a local structure that develops itself even at the expense of the rest. In the end, it kills the rest. Every disease can be seen in informational terms as a reduction or blockage of coherence.

What does this mean in practical terms? It means that we have to search for the fine-tuning of all the elements that make up our organism. Every living species does it instinctively. We are the only species that is disorganizing itself, disconnecting itself from the rest. It seems to me that the challenge of health in an organism is to find the potential for coherence in a sea of incoherence.

The other aspect on this practical level is to fine-tune our connection with the world around us. In the largest perspective the world around us is the cosmos, but our task centers not on the cosmos, but on our most immediate surroundings—on our loved ones, the core elements of our coherence, and the sequentially wider groups and systems in which we are embedded. Trying to fit ourselves in such a way that this entire gigantic cluster we call the web of life becomes sufficiently coherent to maintain itself and evolve. It can't maintain itself in a static way. It either evolves or it devolves. There are always fluctuations. There are always multiple impacts. Therefore we have to be viable enough to evolve in the face of chaos and incoherence. Overcoming them is the practical task of our life. It is to search for health, balance, and dynamic equilibrium, to search for the continuation of our evolution.

Tobias: Beautiful. Beyond cognitive dissonance, beyond anxiety, beyond turmoil, beyond all the collision points. It leads to what I call suppositional possibilities for survival. A diaspora of hopes. Perceiving the universe and all those evolutionary impulses from the time of the Big Bang from the point of view of both astrogeophysics—short pulsar bursts, the birth of the universe some 13.8 billion light years ago, at the very edge of scattered light—as well as from the current evolutionary vantage point of punctuated ecodynamics, the belief in bifurcation points, namely, the traditional theory of punctuated evolutionary forcing that arises on the cusp of all-out crisis. Forcing comes internally, as with greenhouse gases, or externally, as with solar radiation. But what about evolution forcing itself? Vast swathes of abrupt change occurring because of intangible buildups of trends over time. Fascinating, poignant, harrowing, demonstrative. At stake today is nothing less than the sheer survival of everything we know, as sudden changes occur in genotypes or behavior of third generation offspring. This includes minute details that blast their way into our consciousness and seem to change all the rules of environmental and survival etiquette. At stake are our birthrights and those of trillions of other wondrous creatures who live alongside our feeble, meager efforts to be human. We must assert, with humility and jurisprudence and integrity, the being invoked by philosophers from Giambattista Vico to the first Italian society of scientific secrets convened by Girolamo Ruscelli at the height of the Renaissance.

Laszlo: The cusp of the process of evolution is bifurcation, which is danger as well as opportunity. Evolution is not a linear process, it is highly nonlinear: it is interrupted, but it is never stopped. And the opportunity arises at these points of disruption, then we have the opportunity to change. But when this opportunity is perceived consciously, the chance of finding a positive outcome is enhanced. Then we can take the process in our hands (which is to slightly overstate it), and we can foster it and promote the search for a positive outcome— meaning the conditions that smooth the path and increase the probability of an unfolding that favors our existence and our evolution.

Every bifurcation, every cusp (unless it results in a breakdown) will ultimately give rise to the next level of evolution, enabling the recognition that there is a new way of moving forward. But bifurcation can be a fallback as well. It can be a reversal that can take anywhere from seconds to eons to correct.

TOBIAS: And hence, the current Anthropocene, unleashed unilaterally by our species.

LASZLO: If we are able to recognize the process of bifurcation consciously, then it's our opportunity to optimize the chance that it unfolds with minimum risk of fallback and maximum probability of moving forward.

TOBIAS: And in the consideration of life outside our solar system—the area outside our solar system being such a proliferating realm of opportunities, albeit in linear time distant both in our imaginations and in the very technologies we use to apprehend those images—we still have this persistent dialogue among all who consider the rarity of life on Earth against the prospect of organic beings existing elsewhere in the universe an irreversible, statistical likelihood. We've detected signals, and perceived organic potential outside of our system down to the second and square foot, including the frost atop giant sand dunes on Mars. And now with all these new, exobiologically enticing planets in the so-called Goldilocks Zones. But it's of little comfort or solace to those of us here on Earth who are grappling with the life force we are entrusted with. That trust is a compact that turns upon personal commitments to kindness, empathy, to what I would suggest is the solace system, not the solar system.

Nonetheless, the word *opportunity,* to my way of thinking, has huge deployment potential in that one of the foremost dilemmas of our age is this systemic depression and ill health, a malignancy that runs rampant in our veins literally, like the destruction of the Bialowieza Forest, Europe's last Ice Age climax forest. You could break it down further into the very mundane but ferociously serious disease prevalence in humans themselves, who, in a sense, are their own bark beetles, a subject Jung, Freud, Claude Lévi-Strauss, and Bruno Bettelheim have certainly addressed. Our ailments stand out, whether it's diabetes in children or obesity, or any number of the most notorious killers of humans: heart disease, depression, loneliness, suicide among teenage girls in remote corners of Ontario, an epidemic of MDs also committing suicide in the United States, and so on.

So we focus on those things in the concrete sense—a chair is actually a chair, forget Gertrude Stein, Plato, Hegel, or Merleau-Ponty, these are the basics. And when we speak of opportunities, as a rule we don't do so in juxtaposition to the contemplation of the stars or of future life on the moon. We skate on thin ice between simplicity and absurdity, because a human lunar colony would provide us with nothing that we can't provide for ourselves here on Earth. It is a certainty that we will be carrying our own genetic baggage with us no matter where we think we're escaping to, as we have done with every continent here at home. We'll disseminate the proclivities that have always conferred a special privilege in terms of our transcending the very carrying capacities that give us our lives. We ignore those boundaries, as we always have, at our constant and continuous peril. Destruction wherever we go. It is foolhardy to think otherwise.

But if we embrace the opportunities you have just referenced and are enshrining, it seems to me that this cusp, this bifurcation point has every reason to spark a genuine contagion of enthusiasm for nurturance and compassion. We have plenty of compassion and yet it remains in great scarcity on Earth, in a sense. There are no troy ounces of compassion, no futures frenzy to profit from on tolerance or the legacy of Saint Francis on Wall Street. Yet, we see such love in every nurturance setting and event. Not in event horizons, but in a courtyard in seventeenth century Delft, as witnessed by a Vermeer; a picnic as Fragonard would have it or a Fête Champêtre in the hands of such artists as Giorgione or Watteau. In every household where there is a mother and a child, a father and a son, a family unit. Of course, all these units differ from culture to culture. In some units you have four wives and one man. In other units you have two partners, while in others, like the Dutch coffee house renaissance, you have solitaires sitting side by side focused on their computers, isolated from one another but clearly expressing, I would argue, an inherent need to be in communion with others of our species. It often seems to me that Giacometti's sculptures of the thin man and his shadow depict the two on the move seeking companionship. In communitarian units, as in circular Kung villages, everybody knows who it was that sneezed a moment ago, who is having sex with her or his partner, who is fighting, who is sick, who is angry, etc.

The same levels of inherent and intense socialization have been suggested even in terms of suicides off the Golden Gate Bridge in San Francisco, where the vast majority of the hundreds of deaths have occurred on the east side of the bridge facing the city, not on the western side where one looks out into the vast, unfathomable fringes of the infinite and cold ocean, with its enigmatic seamounts that encompass the virtually inaccessible fogbound Farallon Islands. There are any number of kinds of units and configurations here on Earth where such "opportunities" are lost, it seems to me, where we've acted obliviously to all that surrounds us. This same "all," the other, has marked our kind with a unique challenge. I call it a responsibility of compassion, or the burden of compassion. However solitary one's poetical self, the recognition of biodiversity confers a social responsibility on all of us.

All that said, it seems apparent that the laws of nonviolence do not seem to be working, although I think that as Standard Operating Procedures they are basic to human physiology. The biology of kin altruism and group fitness has been well developed so that our predisposition to be kind, tender, and loving is widespread and universal. We've grown up with it. I come from an optimistic persuasion, despite being a Jew with a history of unspeakable horrors meted out to our people. But still, I don't believe as others do that violence is the operational code of our being. I believe, I want to believe, I *do* believe that love and nonviolence are the true mascots of what it takes to behave as a human being, and that we're given to that, as Dame Jane Goodall wrote so beautifully in her preface to your recent book, *The Intelligence of the Cosmos*. And I think that the interference, the dissonance that we encounter blinds us to that opportunity.

LASZLO: That is a danger, but it's not fated.

TOBIAS: I'm hoping against hope that in this generation we can find the words, the expressions, and the examples in ourselves to make manifest a universal language that's accessible to everyone and that offers up no hindrance, no obstruction whatsoever. It may come in unexpected forms, in unexpected ways, but the form in which it comes is irrelevant. And as our density escalates as a species in sheer numbers I think—given the fact we're adding approximately a billion people every twelve years, 150 million net births, 82 million survivors of that birthing ordeal annually to the planet, and 233,000 per day—this becomes an increasingly heavy burden for us. Put that in the context of opportunity. If you are at the World Bank or the UN or a CEO or a young person wandering into the world, the challenge of seeing human overpopulation as an opportunity for tackling the most grievous consumption issue in the history of humanity head-on is almost a deafening roar in one's ears. For instance, Niger, a country with one of the highest poverty levels, has the highest fertility rate in the world. The birth rate for a Niger woman averaged more than 7 children in 2016. We all see that something must be done about the increase of birthrates in many counties at the same time that all of our human consumer habits are multiplied, whether we are rich or poor, on top of our further dilemma of grappling with millions of hungry people already struggling at the edge of defeat.

And I pray that we can communicate this great challenge before us right now. It's bewildering to me. I come back from a journey that I had this morning behind your house and I feel like the crickets last night, their antennae moving a hundred times a second—sexual lures—were reminding me that other species do not exceed their fertility boundaries for more than a few generations. If they do, they go extinct, as spelled out in the boom and bust biological credo that declares sustainable or unsustainable demographics. Humans fail to take note of these warning signs wherein we should be limiting the size of our families. This is a major chord in our evolution, and it is either harmonious or dissonant. That's up to us, to the choices we must make.

And then I observe before me a bumper-to-bumper, self-destructive community of insane, trapped commuters, such as those on the 405, as it is famously called, or the "10" in Los Angeles, and those closed ecological systems known as freeways. Hot tarmac. Road rages.

To be immersed in the flow atop that unspeakably morbid tarmac invites consideration of all those individuals lost in their automobiles. People are gunning it on such roads, anxious to get somewhere. To escape the chaos. But where are they going? Do they know where they are going, in the big picture sense? They are ambassadors of a system that has broken down. A symbol of our species no matter how much consciousness may have escaped the many stellar implosions, supernovas, black holes and all that energy associated with dark background matter particles. So, I ask again: do we know where we are going, do we know what we are doing? Or are we driving these roads by rote, thousands of such hours between our fleeting birth and death? Can we step out of that freeway system and sit back and really see beyond the morass, truly embracing those things that will have a deep resonance long after we're gone? Can we, *are* we fulfilling some kind of promise, a contract with the universe, the cosmos herself, that is vested in us and makes our humanity special—out on a freeway? All of these questions plague me. But they also give me great joy at the same time or I wouldn't bother trying to express them.

LASZLO: The way forward, if we find it, is by recognizing that joy, love, empathy, and sustainability are not arbitrary acts, but acts rooted at the core of the cosmos. They are an expression of the reality that is in us and around us. This recognition has to be based on the insight that life is not a chance phenomenon, not something that is rare and just happens to be here because all the physical, biological, and ecological conditions for it to emerge are fulfilled on this planet—not something that is extremely rare in the universe. Yet this was the dominant conception until a few years ago. Lately it is becoming recognized that there are instantaneous entanglements, nonlocal configurations that create the structures that are basic to life. We now find that organic molecules are coming about in the course of the chemical evolution of stars. This is completely improbable in terms of the dominant paradigm. Yet using space telescopes we find organic molecules, the basic building blocks of life, in the interstellar dust that eventually gives birth to planets. It seems incredible. How did they get there?

Life is a universal force. It expresses the same trend that is expressed in the formation of galaxies and the formation of atoms. It is the force that creates relationships, producing higher and higher level structures. If we take account of this, we will know that it is in us as well. Love, solidarity, empathy—these are basic expressions of the evolutionary impetus encoded in the formation of atoms, the evolution of organisms and of galaxies. It is there in us as well. The recognition that we express the basic evolutionary trend can give us direction; it can give us confidence. Because in us is all the wisdom that we need to exist, to evolve, and to flourish.

TOBIAS: That's a great insight.

LASZLO: Finding the information that forms the universe is not something we just dream up. We have to search for it, we have to unearth it, digging below the avalanche of irrelevant superficial information. It is there in us—our bodies could not survive without it.

We discover this evolutionary impetus in previously quite unthinkable ways and places. I mentioned organic molecules. Lately they discovered that DNA strands can spontaneously recognize one another across distance, though mainstream science does not understand how that would be possible. This kind of phenomenon indicates the existence of nonlocality, which means that there is information, coherence, and consistency across intervening space. Space doesn't separate things, it joins things.

TOBIAS: Many years ago Rupert Sheldrake started talking about dogs knowing fifteen minutes prior to when the Ervins and Caritas of the world are coming home to their four-legged friends. I am thinking of the Australian Maremma sheepdog Oddball, the hero who saved a colony of native penguins from foxes in Middle Island, Australia. This central Italian breed of guard dog knows instinctively when hostilities are emergent. Their trans-species altruism expands the circles and layers of family beyond normal meanings and gives us a very down-to-earth chance to recognize the physics in belief and expression patterns, and particularly in nurturance. To see it between species is very special. I remember the early studies on the electromagnetic resonance—sensing through invisible borders all manner of phenomena—and how that's possible, and then the various tests that were done. It's easy to test at that simple level because you have something on the order of 60 million dogs living with 320 million people just in the United States, so there are testable daily realities that transcend conventional illogic. We now see it to be the case. But we don't understand it.

LASZLO: The basic factor is always this entanglement, this nonlocality that was first discovered on the quantum level but that turns out to exist on all levels. It turns out that the coherence of the living system is ensured by this nonlocal connection. Just think: Information flows in the body by biochemical means at the rate of signals traveling along nerve pathways at twenty meters per second. With this speed, it would be impossible to manage the billions of chemical reactions that occur in the body, millions of which occur every second in every cell.

TOBIAS: You remember the famous challenge, if you had to choose between taking over your body and maintaining all the controls or commandeering the immensely complex controls of a plunging 747 in a dark hurricane, out of fuel, and with all four engines dead, one would still be a fool to think her or his chances of controlling one's own body guaranteed better luck, because they don't. We'd be dead. Take the plane rather than the helm of your body. I don't care how perfect a yogic master you think yourself to be. The involuntary musculature, all of the synapses, trillions of connections, billions of letters in our DNA, and it never ceases to enthrall me (and not without a smile) in the ethological realm, the interspecies communication realm, the semiosphere, where we are looking at the semiotic, the signals between species. I think one of the most exciting examples of this is root exudation. So exciting my English skips grammar, as you have no doubt noticed as it leaps from topic to topic, subject and noun to adverb and superlative. This is because I am utterly amazed by the anthrozoological connections that our species has—certainly since the Mesolithic—ignored. We have refused to look all of our fellow denizens of this earth in the eye in terms of real-life communication.

But back to the root exudation: we now know that there's molecular communication going on in the soil between rhizomes, the rhizosphere, and the larger plant species above. We have good reasons to believe that nitrogen fixation, for example, is a communication process. Equally enthralling is the prospect that evapotranspiration in the stomata of every plant (the so-called FAO-56 method of water vaporization from every living, green, chlorophyll-emitting substance on this planet) is also a process of communication. Put that in the context that every act of communication is a reciprocal act of translation (as the brilliant litterateur and philosopher George Steiner once wrote, I think in his masterpiece—one of many—*After Babel: Aspects of Language and Translation*) and you have a level of connectivity in every earthly domain that is simply astonishing and wonderful. There are transmitting entities and receiving ends, and all of them are alive. It's simply thrilling. I can hardly sleep at night, it is so amazing to me. Hence, my chronic condition since childhood of insomnia. I am so absorbed by it all. The transmission is transferring information and so I agree with your theories 100 percent.

What is so amazing and exciting to me is the fact that with the 100 million plus species we are now aware of—not including the world of viruses by the way, which exponentially increases that number—the amount of information being conveyed over these distances is staggering. When I come back to the present and the concrete reality of standing there by the Cecina River behind your home before dawn and listening to the vast symphony of life—pure and simple acoustical cognizance on my part—I didn't . . . I couldn't even begin to grasp the green Chartres Cathedral within the Amazon within the microcosmic whorl of that river. It was as though I were walking through the rings of Saturn for the first time. I was so enthralled that it left me speechless and in tears—that riparian truth of Tuscany, known for the most densely forested patches in the whole country. But this was the untouched riverbank, macchia vegetation as it's called, perfectly inaccessible, with glossy leaflets under the moonlight, double reflections off the creek, and all kinds of evergreen, knee holly, and myrtle. An odd enormous chestnut tree, beech, popular, and various species of the swamp or *padule*, as they say here. And I kept hoping to see badgers or one of the remarkable and diverse species of European polecat, or maybe an otter or beaver, sleeping geese, nightingales and nocturnal predators on the lookout. The bright eyes everywhere.

And that's just in a little creek in Italy, north of Rome, south of Pisa, a tiny little creek with its own problems of various pollutants

—chromium for one, but also nonorganic point and nonpoint agricultural effluents, and so on. And if I can still feel this way after all these years of doing these things, these little personal expeditions into the wild, going out there—the thought of getting jaded or of it ever becoming monotonous seems impossible. The joy that consumes me is contagious within myself to the extent that I am . . . well, let's just admit to ourselves that these wild epiphanies constitute an ineffable experience. It is everlasting. And if I could only share it, if I could only deliver that sense of total divine mystery . . . ahhh, it's pure bliss. And I can't imagine that a second passes in any child's life that their heart isn't awakening to such feelings, or that the whole placental process, the nurturing process, biologically, biochemically, isn't about that. Gestation and incubation is at the progenerating core of all life in mammals, birds, reptiles, amphibians. This profound mystery behind your house then carries over to the twenty-two billion bacteria in our armpits, the seven million follicle mites enjoying life in our eyebrows, the *E. coli* in our gut.

Just a few years ago a Russian scientist discovered an immortal bacterium in Siberia whose genetics are now being studied at Moscow State University's geocryology department. I'm speaking of Anatoli Brouchkov's remarkable discovery around 2009 of the 3.5 million year old ("eternity") species *Bacillus f.* A Jain monk would never dare hurt those microbes, nor even shake someone's hand for fear the friction could cause injury, or "hinsa" (harm, versus ahimsa, non-harm or nonviolence—the essence of Mahavira's message). This was the key "mahavrata" or major vow in Jainism, which so influenced the entire philosophy and "be the change" message delivered cathartically to a nation by Gandhi. Even to walk on the grass as we did yesterday is an injustice to that grass. That sense of injustice may be more widespread than we usually are apt to credit in people. That we feel these injunctions—like not walking on the grass—is a sensation, however rarified and quietly guiding, that may well have informed the best of every belief system.

Our sense of hesitancy and respect defines much ethical and indigenous spirituality, dating back several hundred thousand years, if not more. I was thinking of the Mbuti Pygmies staring up at the heavens and the earliest Australasian peoples meditating on, say, the unbridled beauty of the constellation Cassiopeia or the first comet. We know of such primeval astronomical acumen from petroglyphs and research by paleontological and linguistic explorers like the great André Leroi-Gourhan, and then André Malraux, the discoveries at Chauvet in the mid-1990s, and so many who have focused on the sheer majesty of this universe. I just don't understand why we don't seize that opportunity we referred to earlier. Why we make it so difficult to get it right. Especially when it's there in us, as you say. We embody it. Why is our species making it all so difficult? This is more than an annoyance to me.

LASZLO: Having it in us is one thing, recognizing it and making use of it is yet another thing.

TOBIAS: I think back to that genre which is conveniently thought of as the anthrozoological gaze. Cervantes was one of its practitioners when he nurtured a dialogue in "El Coloquio de los Perros," "The Colloquy (or Dialogue) of the Dogs," much like you and I are presently engaged in. Cervantes' dialogue was between two dogs, Scipio and Berganza. In his foreword to a 1968 republication of the work in London for S. Bladon, with the assistance of the Huntington Library in San Marino, California, Hal Stebbins articulated a very fine point, applicable to all of nature and humanity's odd role within it. Namely, that all art essentially is a rearrangement of previous perceptions. It is the same every time we look to understand the relationship between the sun and a sunflower, or between that gorgeous Cecina River Valley and this lovely Villa Franatoni up above it, wherein two humans are having a little conversation.

LASZLO: The answer lies in the old saying that is rediscovered and more and more recalled today: that as a modern and healthily skeptical person you believe what you see. But you also see what you believe—and only or mostly just what you believe. Because what you don't believe, you find difficult even to see. There are a lot of important things we don't believe in, and so we don't really see. Our beliefs have gone astray. We went off in a direction of thinking that all these things we talk about are just imagination and poetry, that they are not there in nature and have no roots in reality. As long as we think like this, adopt this mechanistic and reductionist worldview, a lot of historical developments seem like accidents. Yet they were part of the unfolding of trends that are coded at the very roots of our being.

It is useful to remember Galileo's insight. He divided what we experience into two categories—those with primary qualities and others with secondary qualities. Primary qualities are objectively present in nature, whereas secondary qualities are what we add to our experience. Secondary qualities include our ethics, our belief systems, and our aspirations. In Galileo's time, these had to be handed to the church because the church was the highest spiritual authority. But primary qualities are present in nature, and the church agreed that these can be investigated by science.

With the worldview enunciated by Descartes, this has led to a fundamental split in our view of the world, to "dualism." Science is investigating physical, measurable qualities. When we find phenomena that cannot be measured with physical instruments, and when we believe that they are real, we end up with a dualist world. The part of it investigated in science is a despiritualized material-mechanistic universe. In this universe, everything that is real is measurable, and everything real can be manipulated . . . we only need money and power. With this view, we are led down the wrong path. We are convinced that we can use money and power to get the things as we want. We can pursue our self-interests—our selfish interests. Our interests express a basic law of this mechanistic universe; the stronger wins because the stronger has more power to manipulate the universe.

TOBIAS: If Darwin had had more time he might have repented, who knows. I mean Spinoza, who was more or less excoriated by the Dutch Orthodox Jewry of his time, came up with an ethical system that was breathtaking and equally misunderstood in his era. His philosophical ground rules—his definition of God as the collective energies of the universe—were deemed to be heretical. And Descartes of course, I mean I would amend his "Cogito" by saying "I feel, therefore I am," or *sentio ergo sum*, and apply its perdurable veracity to those ineffable qualities as opposed to this brute force that is the megatonnage of human violence, basically, and of this mechanistic approach to life which, thank God, we're slowly growing out of.

But when I read about certain Iranians burying massive numbers of dogs alive, I truly rage inside and have no clue how to think about redeeming our species. That's just one of countless examples that smacks me over the head every day I wake up and roam through the social media that conspires to overwhelm us. We've "outgrown," tragically, our better selves in so many guises. The hope infiltrates the fact that even those who have outgrown their innocence are still vulnerable to change. We still come across true windows of our real humanity every day. Consider the work the late John Nash did with the remarkable Tasaday tribe of Mindanao in the southern Philippines back in the late 1960s. The Tasaday are a peaceful people, not Cartesians. Or think of the Toda of South India in Tamil Nadu. They are among the last vegetarian tribes on the planet, all 1,200 or so of them. They are not some ethnographic anomaly, but a true source of information and an example of what is possible among our species. They worship their river buffalo and leave virtually no footprint whatsoever.

Were today's urban-driven societies to somehow find ways to emulate such tribal last calls, our days and nights would be focused on relaxation, swinging in hammocks, adoring our babies, spending very little time, relatively, gathering our food sources, let alone hunting. We would be essentially the way we were 75,000 years ago. We numbered at that time probably fewer than 100,000 individuals. And the conditions for our existence . . . amazing. It's very difficult to analyze. We can romanticize it, but in fact we have sufficient data to interpolate a modus operandi of societal existence which, I must confess, looks rather attractive to me. Evidence suggests that life 75,000 years ago anywhere on Earth for humanity was perfect. Yes, there were challenges; yes, the life expectancy was less than thirty years of age. You got the odd individual who would live to be eighty, but that was most certainly not the norm. Even by the time of Christ's existence life expectancy was something like 32; there were 275 million people approximately on the planet in the year one AD. There were already cities of a million in size.

But now, to speak of these things in terms of the past is more than a harmful anachronism in the sense that it hurts to even think about it. As David Rothenberg has said, it's painful to think about it, but I go back to the word "opportunity." I sense that with the scientific breakthroughs that are happening at breakneck speed, we have the daily or hourly avalanche of data concerning these revelatory messages. They are being delivered to us by our very curiosities, our voracious appetites for technologies, for seeking, for the generalist who is struggling, who is caught up in what in Asiatic traditions is usually regarded as "the whorl of existence," that miasma, that struggle. The opportunity is there in all of the surveys, for example, conducted from country to country, in schools, in cities. The majority of human beings are hopeful.

You go to a refugee camp in the South Sudan or Afghanistan, or head downtown in Beirut having driven westward over the mountains from Damascus, as I have, and you ask people who don't even have a glass of water in a day if they have hope. And I've questioned people like this throughout the world who are in very tough conditions, and they say, "Yes, I believe within a year or so we may have electricity," or "My dream is to get a refrigerator, and I have every expectation that we'll get it." This latter example came from a woman in her hovel in the so-called Nehru Slum in New Delhi some years ago. Or, "I may not have shoes, but I already have a three-year-old iPhone." It's crazy, but that's the world we inhabit. Humanity is poignantly bizarre, distracted, distanced from the creation. We have this ability to smile in the face of horror. That's a curious phenomenon, perhaps even a key to our salvation as a species. The fact that we can cope in spite of our species' ungainliness and trillions of homicides against the biosphere. Maybe we are insane to be capable of such cruelty while wearing a smile. And maybe that is the norm. But I don't think anything can stop us, which is both good and bad.

LASZLO: Except ourselves.

TOBIAS: Well, yes.

LASZLO: That is the problem we started talking about this morning; the problem of a process we have engendered without being conscious of it, and without wanting it. In our world the force that is behind the evolutionary trend is not expressed in a way that could ensure its unfolding. As a result, our core being, our sense of being and flourishing, the smile, the joy, the recognition of oneness and of love in the world, is buried under the weight of signs of impending disaster—the acidification of the oceans, the heating up of the atmosphere, the extinction of species, and the violence that surfaces in the conflict between societies and cultures. All these processes are unfolding simultaneously, and they make the outcome of the evolutionary process undecided.

Yet I believe that this process harbors an element of self-determination. Given time, it will be decided by the fluctuations, the small but cumulative processes of change that come about on the way to a major bifurcation. These small but crucial fluctuations are up to us to produce a smile on the face of a child in a favela in South America and in a refugee camp in Asia. That smile is a fluctuation that can counterbalance the processes that lead to the extinction of species and the other catastrophic consequences of an unchecked bifurcation.

DAY FOUR

Morning, by the River

TOBIAS: Speaking of unchecked bifurcation, as we did last night, I've seen it in a Bangladeshi refugee camp. Fifty thousand individuals, mostly women and children, who had been herded like all those poor goats of the world, tossed into the backs of trucks during the mid-1970s in Dhaka, and taken to a camp in a town called Tongi, where I watched as they starved to death. The women asked me to photograph them and their children for some horrifying posterity or else from the hope that posterity might see in those unbelievable images of extraordinarily courageous women a chance to do things differently in the future. To fight back with everything you've got against tyranny. They were indelible images that plague me to this day, particularly in light of the reality that some 800 million young people and their mothers are hungry and that malnutrition persists into the twenty-first century. I was there to see that horror in Tongi. I've published some of the images of these women. Most of them would perish within weeks of those tragic few days in my own life.

LASZLO: Like in the Nazi concentration camps.

TOBIAS: Yes. And I think about those children drawing butterflies and writing poetry at the Terezin Concentration Camp north of Prague. It all leads me to a newly emergent category—not a law of thermodynamics, but engaged ethics, which renders the human condition altogether rich with sweeping and magnificent opportunities for planetary redemption. And in uttering these words I'm mindful of what I have explored most of my life in tracking what my muse and my life partner, Jane, and I have together termed "the cartography of pain points"—namely those flash points where the worst, unimaginably horrifying, and cruel aggregates of suffering are meted out by our kind to those others. The crucibles of callous, human indifference: A turkey in Arkansas, a coral reef off the coast of Sri Lanka, a Kapok tree in Brazil, a mountain gorilla in Rwanda, a slaughterhouse in Iowa, or the camp I just referenced in 1975 in Bangladesh.

The question that I'm really asking you, Ervin, hinges on the "engaged ethics" (my use of the phrase) here on Earth that may be connected by vibrational energy to the "intelligence of the cosmos," as you have named it in your newest book. But it also depends on an ethic that will play out in the hands of either feeling or indifferent human beings, as our entire species segues from the opportunities of expressing evolution in ourselves to the very edge of the bifurcation point—that punctuation point of evolutionary change. It is there that we choose whether or not to convey the information that we are coded with in a manner that is self-aware and authentic (as Jean-Paul Sartre would use the word). There are distinct genetic advantages to doing so, to communicating our epiphanies and inspiration as opposed to remaining silent, adrift, and fated, more than likely, to premature extinction. I'm speaking of a shift towards totally authentic consciousness, as opposed to some superficial performance.

I think back to my many months across the western Tibetan Plateau visiting monks who for six hours per day, starting at 3:42 in the morning, tried to even come close to the depth of "Om mani padme hum." Six hours of repetition, continuous and trance-like (the Ladakhi translation being, "I invoke the path of universality so that the jeweline luminosity of the immortal mind be unfolded within the depths of lotus-centered consciousness and I be wafted by the ecstasy of breaking through all bonds and horizons") only to discover that it's not a trance at all. Instead it's an acutely fine-tuned oracle inchoate, not unlike perhaps your early performances of Liszt on the piano. We spoke of that absolute precision that is a choreography in perfect, Pythagorean harmony.

But if it's a Jingdezhen porcelain made for the seventeenth century Japanese market, it must contain a flaw so as to fit in with the concept of *wabi-sabi,* which began to gain popularity at the time, and make it truly human and desirable (a trait that was somewhat counter to traditional perfectionism amongst Chinese aesthetes.) Those Tibetan monks I got to know were not in a trance but a state of pure being. They can snap their fingers in the middle of their mantra and go let a mouse outside. It's not some pseudo-omnipotence because they are not in their Nirvana—they are working towards it. But they are engaged.

And I suppose that's what struck me when Jean-Paul Sartre turned down the Nobel Prize for Literature. In some ways emulating Zola, he was saying, as I interpret it, that if we are not engaged, then all of our philosophical meanderings, our being and nothingness, our dialectics, our synthesis, antithesis, and thesis—all of this is nonsense, a form of hypocrisy in the end. Especially if we don't engage in this life with our friends, with those people across the street, with the other, with mountain gorillas and turkeys, examples I use because our species slaughters them. But it could equally be a person of color, a member of the LGBTQ community, the Native Americans that we massacred in the United States, the slaves that we shipped from Africa to Georgia, and so on and so on. The Kristallnachts of history. It seems to me that if we don't embrace ethics, we have lost the opportunity to even contemplate the stars or evolution, or truly extol the virtues of a Monteverdi. And evolution has granted us, as I think we have been establishing here, an opportunity we are missing.

LASZLO: Yes, ethics has to be rescued from the philosophical framework of purely cognitive activity. It has to become *ethos*. It's something you embrace, something you live by. You don't have to cognize it or comprehend it, you have to live it. And that's the great question. Can an intellectual insight lead to such a shift? People who come to that kind of insight are themselves the exceptions, of which you yourself are one. They are surrounded, however, by people who are too intellectually stymied to be able to move beyond their own cognitive frameworks.

TOBIAS: Nikos Kazantzakis always called himself a "pen pusher," yet this was the same Nikos who, after forty days and nights on Mount Athos accompanied by his friend the great Greek poet Angelos Sikelianos, descended in the winter of 1915 or 1916 to a household to have tea and found in the courtyard beneath the mountain an almond tree. He wrote in his autobiography, *Report to Greco* (Greco meaning the painter El Greco, who harkened from Crete and whose real name was Theotocopoulos), "I stepped up to the almond tree and said, 'Sister, speak to me of God,' and the almond tree blossomed." This was a man who translated all the great masters in a dozen languages; he wrote dictionaries, over fifty books in the ancient and modern Greek—a true genius. And here, this humble explorer who just spent forty days on an ascetic retreat, sees this bare almond tree in the winter, this blossoming tree. This is an ethical human being who inculcates within himself the fullest manifestation of the human heart and its perceptive capabilities.

LASZLO: Yes. This is perhaps an evolutionary mutation. It is emerging here and there. Does it emerge fast enough, does it spread fast enough to make a real impact, to make that crucial difference that we need to move from one branch of that tree to another, from a dead branch to a viable one? That is the question. No doubt it will be answered. We have to buy time.

TOBIAS: Buy time?

LASZLO: Yes. We have to do our best to buy enough time so that these evolutionary processes, which are ultimately processes of recognition of who we are and what the world is, unfold sufficiently to support and motivate the new ethos.

TOBIAS: I was impressed that in the very beginning of one of your books you referenced the Jains. I refer here to what I call the "biochemistry of the nigodas," those atomistic components of the jiva, the immortal souls in all being. This is no value judgment but the embracing of a striking and immemorial possibility, namely what the *Acharanga Sutra*, very early compilations based on the teachings of Mahavira characterized chiefly as the interdependence of all living beings and their mutual collaboration, in other words, beings caring for one another. In a more modern context, that would translate well into the aforementioned biophilia, that accrual of cumulative instincts and persuasions, of predilections, likes and loves, friendships and co-sympathies that all come down to organisms enjoying one another's company; different species commingling, as right out of the Book of Isaiah. Such life-fostering juxtapositions in the history of ethics, philosophy, art, and spiritualism help us to wrap our minds around all of these contradictions we have been evincing. A tool towards our very salvation. It's impressive to me that like Democritus and the other atomists in pre-Socratic Greece, even before that time, with thinkers like Thales, who attributed a life force throughout the seas, you had in Asia this multiplicity of traditions—Jain, pre-Buddhist, or Bonpo in Tibet, the early Indus River Valley civilizations, which were sworn to loyalty to this belief that every dew drop, every dust particle, every mote of light contained life. It was alien of course in Western history until quite recently.

LASZLO: Panpsychism, which was there in one form or another all through history. What you quote are symbolic expressions of it.

TOBIAS: I call it the future of biophilia. Evidence-based optimism by way of that manuscript of nature which our own species has struggled for 330,000 years or more to read, offering up to this phase in our consciousness a penetrating glimpse upon a level of biophilia, of compassion in the wilderness that comports with our greatest scientific, aesthetic, and spiritual arousals. What do you see in regard to the future of biophilia? Biophilia, of course, referencing that innate, collaborative zeal that we see so pronounced for instance in that famous *National Geographic* cover story of the polar bear in Churchill, Canada. The polar bear, which generally came into the town to raid garbage cans, encountered a husky chained to a house with a long leash. The photographer who witnessed this expected the worse, but the polar bear and the dog started to play. They played for fifteen minutes and it became a cover story and went viral around the world. It was one of the most popular issues of *National Geographic* in history. And that's biophilia—this innate affiliation that life has with itself.

Of course, the worst species demarcation zones that conventional biology has imposed on the natural world to define and conquer it, to master the language and the lexicon, is becoming quickly obsolete. Taxonomy is becoming a nonstarter. In fact, it is dangerous to designate species and subspecies because that involves legal conflict in terms of US Fish and Wildlife type agencies throughout the world then being put on public notice to make biological opinions and determinations as to whether or not to safeguard those taxons. That in turn makes instant enemies of developers, which in turn takes lawyers and judges and juries and taxpayer money. The system of protection in other words is utterly backwards, with disincentives rather than science driving the entire collective bargaining process. And this is just wrong. Legislation should never be held hostage to cultural expediency. Carl Linnaeus was the grand chess master of nomenclature in his *Systema Naturae,* first published in 1735, and it continues to this day. That man had more influence than almost anybody in the history of science, in opposition, of course, with the current US political administration which banished all science and for the first time in many generations has ousted a science advisor in the White House.

The implication of these issues in terms of everyday protection—what Jane and I went to great lengths in our book *The Metaphysics of Protection* to lay out—is the fact that science still clings to Linneaus' nomenclature. Since his tenth edition of the two-volume *Systema,* published in 1758 in Sweden, we haven't changed. It is the lingua franca of every biological course taught in every university around the world, and the genus and the species have become our divining rods of how we see life. But the ineffable poetry of nature has nothing to do with taxonomy, of course. And in spite of his insistent rules, Linneaus first applied his taxonomy to botany. And he was unabashedly in love with life and its diversity and felt compelled to rationalize it all and place some sort of order on it. He was so enthusiastic he even went to great lengths to incorporate chocolate—the wild source of chocolate (*Theobroma cacao*)—and the diminutive evergreen tree in the Malvaceae family. I've tasted one of its large beans where it had fallen from the tree onto the ground in Ecuador's northeastern Amazon, and it is truly delicious.

The speculative insight I am trying to catalyze in you is that when you look into the future through the eyes of Ervin Laszlo, do you see humans relating to nature in a way that might be radically different than what we are seeing today? What we see today in Jane's and my "cartography of pain points" is a mass consumption of other species, brutalization of other life-forms, the equivalent throughout the biosphere of what José Ortega y Gasset described as the dehumanization of the masses. We're doing this in every slaughterhouse and in every concentrated agricultural feeding lot with 400,000 chickens on average confined per lot where they can't even turn around. And for us, concentrated in these inhumane megacities, what do you see in our future?

LASZLO: That's really asking whether I or anyone can see the future itself.

TOBIAS: But you do see it?

LASZLO: There are many possible futures. Which of them comes about depends on how we deal with the current unsustainability and the change or reversal of our current direction. If we get to the point of a massive human extinction or population collapse, the remaining humans will not necessarily see which way to go. Will they have learned from the past and developed a sustainable ethos and harmony with the rest of the biosphere, or will they perceive other species as competitors they have to fight for the resources that remain? If we continue unchanged as we are now, there will be a squeeze forever more rare resources and for livable, habitable space. If we are lucky—or wise—the human attitude towards other species will develop in the co-evolutionary perspective, which will be one of sharing and of embracing a synergistic perspective toward others. And if we are not lucky or wise, it will develop as an oppositional relationship, of fighting and competition. The big question will be, as it is already today, competition or cooperation. Competition will bring a linearly increasing series of difficulties, but cooperation could develop as a rapid, nonlinear shift. The development could be very fast. The choice of perspectives is fundamental. We could shift to a new mentality, a mentality of living in a kind of Garden of Eden where we share our habitat with one another and realize that we can live and thrive together. This possibility is still open. I don't think that we can forecast which of these futures will be realized. All we can do is to stack the cards in favor of a positive outcome, toward the realization of a co-evolutionary perspective.

TOBIAS: Do you have a vision of the Peaceable Kingdom, as the Prophet Isaiah posited, where the lion and the lamb and the wolf lie down together?

LASZLO: The Peaceable Kingdom gives the conditions for life, for living and striving to live together with other species around us. But today it is up to us to provide the means for all forms of life to exist and to evolve. How can we do this, and with what means? One of these means is the love and oneness we can feel with each other and with other species. We must come to the insight that it is in our interest to enable other species to live and to thrive because we can't exist much less thrive without them. This feeling of wholeness and oneness determines how we relate to other people and other species. The love for all forms of life is the love for the human species, and the love for the human species is the love for ourselves. This love means that we no longer exist in the exclusive "me or you" way, but in the inclusive "you and me" way, where you and me together are one.

TOBIAS: Martin Buber.

LASZLO: From me to we—from I and Thou to I together with Thou.

TOBIAS: Yes. I actually studied from one of Buber's chief disciples, a philosopher at the University of Tel Aviv nearly forty-five years ago. On a related tangent, it's interesting that the rise of natural capitalism movements—I/Thou economics, if you will—has entranced Wall Street to a point. We saw something on the order of $3.8 trillion in socially responsible investments last year. Young philanthropists are getting the wisdom of NGOs and of giving back to the world à la the Clinton Initiative. On a certain level one could characterize that as a type of biophilia. I think that one of the most striking examples I've experienced of this so-called biophilia was the time I was dropped off a boat in the Sea of Cortez, eight miles out, during a film I was helping to produce on whale sharks. Totally alone, naked, treading water in the ocean, two ten-ton whale sharks, male and female, came up on either side of me, their eyes the size of my head and coming within inches of my own eyes. Had either one of them so much as pivoted his or her body suddenly I would have been instantly mushed. But instead, I communed with them for I don't know how many minutes, possibly even half an hour as they moved quietly through the sea, in an absolute séance of solidarity, emotional connection, total unabating curiosity, and mutual happiness. I have never been so naked, so connected, so out there. These were behemoths the size of school buses, clearly enjoying my company as I relished theirs. It was one of the grandest highlights of my minuscule little existence on this planet.

That film was not about my few moments of tranquility, but rather the horrifying sale of whale shark fins for soup in Taiwan and China, a practice now outlawed in many countries. At the time it was being principally carried out by poachers in the Philippines—just a dozen or so families—as well as others off the coasts of India and Sri Lanka. The film team tracked the poaching industry, eventually convincing the government of the Philippines to render such business illegal and attempt to turn those same poachers into ecotourism guides for people wanting to swim with whale sharks as is now done in other parts of the world. I can't comment on the success of the Philippine government's efforts to that effect, however.

This was back in the late 1990s, at a time when whale shark migration throughout the world was still a mystery, and the role of the blooming of coral polyps as part of the critical whale shark migration cycle was still largely not understood. The state of endangerment of the whale sharks themselves was also not well known. Much more has been learned since then, but for me personally that experience out there in the Sea of Cortez was not unlike landing on an alien world and discovering two friends who just happened to be members of a group of gorgeous beings that constitutes the largest and most mysterious fish species on Earth. In fact, it's one of numerous remarkable experiences I have had on Earth, including a night I spent amid an entire group of hippos bathing in a lake beneath the distant Mount Kilimanjaro. I sat on the side of the lake with no flashlight, just the bright countenance of stars overhead, dangling my legs into the muddy water, utterly at peace with all this commotion amongst at least thirty hippos.

Mind you, I had been warned by my host, the owner of this property, not to go anywhere near that lake, which was about a quarter mile from my hut. But I at once went there anyway, completely at peace with the notion of being with the hippos. And again, it was true communion: biophilia. They sensed my vibes, I'm certain of it. And it was biblical, a kind of Peaceable African Kingdom. At one point some lions came by and must have thought me crazy to be out there with the hippos who, along with the Cape buffalo, zebras, as well as green and black mambas, are thought of as rather dangerous to humans.

Well, it was pure paradise. And if somehow Gaia, or whatever other principles we attribute to the story of life on Earth, that tired rhetoric of might versus right, were to test a theory of nonviolence, the whale sharks and those hippos were living testimony to the fact that planet Earth has lodged in her largest herbivores the most compelling narrative of that parable. That evolution, in all of its vibrations and nuances, material forms and fantastic manifestations, is actually about a remarkable standard operation procedure called gentleness. If the cosmos, as represented here on this planet, far off in a virtually unknown system (compared with the entire expanse of space) can engender such nonviolence in forty thousand cubic board feet of a giant sequoia, or of a fifty-mile patch of moss in Wisconsin, and ton upon ton in such creatures as the remaining ten to twenty-five thousand blue whales, well, what a story to tell. A true story. A poignant one that is critical of our times and at the very heart of biophilia and all that it implies for the future of humanity's role in the biosphere going forward.

Of course, I suppose the skeptic could argue that there is violence on Mount Everest, or Sagarmatha, as it is called in Nepali, or Chomolungma in Tibetan. Avalanches occur frequently and inadvertently kill dozens of people. But as my old friend and hiking companion, the late Tenzing Norgay, used to tell me, he considered the high Himalayas sacred and had very mixed feelings about being the first atop the highest mountain in the world. As you know, Bhutan to this day does not allow climbers to climb to any Bhutanese summits over 6,000 meters because of the sacrality attributed to those peaks. It's very much in keeping with so much that is Bhutanese.

In Tenzing's case, he was like a kid. He simply loved running up mountains, and I can attest to it because he and I crossed half of Sikkim together one winter. I studied his gait, how he moved, the birds he loved, and how he handled an encounter with a Tibetan blue bear we had one night. So many things about him. He was a great ambassador for nonviolence, for biophilia and all life-forms, both sentient and non-sentient. This raises certain aspects of physics into the equation you would appreciate because, like ancient Taoists, Tenzing also had deep Buddhist feelings towards what would ordinarily be described as nonliving things. And Mount Everest was one of them.

To conclude, those whale sharks in Mexico and hippos in Kenya perhaps thought of me as a poor little orphan alone in the water. They protected me, or ignored me. At one point the whale sharks nudged me, and then both got so close that my naked body was literally being conducted through the water by their two bodies—I'm guessing thirty to forty thousand pounds of fish helping me aloft in that foreign warm seawater. Then eventually the boat and its occupants, my colleagues, reappeared on the horizon and came closer and closer and finally took me aboard. These were transcendent, overwhelmingly emotional encounters for which I will remain forever grateful. They are lessons about life that pass through us and are indelible, ineffable, immortal. I was testing a hypothesis. Now I know it is a proved theory. Perhaps these enormous beings simply felt sorry for me, which would prove the "feeling sorry for another" theory, but I prefer to think of it as the proven theory of loving thy neighbor. It works in a true, unconditional, evolutionary context. It works, Ervin.

LASZLO: On this topic of subconscious feeling of connection, I had an interesting and instructive experience a few years ago. I have a friend, a wonderful brain scientist, Professor Haffelder of Stuttgart, Germany. He does experiments where he measures the coherence of the brain in reference to EEG patterns. One day while I was in Stuttgart he put electrodes on my head and measured how my brain responded to various sounds, from Mozart string quartets to natural sounds. Then he played back the recording of the sounds to see how I reacted to the corresponding waves. My reaction to the Mozart was predictable. But there was a surprise: At one point the alpha and the theta waves, the deepest-level EEG waves of the brain, suddenly nearly exploded. I didn't know what had happened at that point. I didn't feel anything in particular. Yet my attention was highly aroused.

It turned out that the sound I was hearing at that point was the sound of a whale mother calling her offspring, warning of danger. I did not know that—I only heard the drawn-out song of a whale. But my brain registered it and responded to it. The point I am making is that we have many sensibilities at our disposal, sensibilities to other species and to nature, many more than we know consciously.

TOBIAS: Exactly. This cannot be overstated, I think. That's why I mentioned the rhizosphere, that realm of communication among plants celebrated in such books as *The Secret Life of Plants,* and *The Secret Life of Trees,* and *The Hidden Life of Trees.* You have conservative Department of Agriculture scientists putting a lot of money in research into communication between roots in order to expedite and accelerate growth, much in the Norman Borlaug school of the Green Revolution, and all the others—the Rockefeller Institute in the Philippines, the rice hybridizations, which would lead to the Codex Alimentarius Commission of the UN, and all of this eruption of controversy over GMOs (genetically modified organisms). US Department of Agriculture specialists are excited by the prospect that these roots that have always been thought to be mindless are communicating in their own fashion, in their own way, and we can tap into it for purposes of exploitation (what else is new?), but also for purposes—

LASZLO: Of understanding.

TOBIAS: Of understanding, that's right. As not only fascinating. It is encouraging. It does lend confidence to the whole impetus and expectation that we might just arrive by a very unexpected pathway at, as Huxley put it, that door of perception into a new world where we truly understand that the world is not flat, and that biophilia really is as readily feeling as, say, Rachmaninoff's famed and oh-so luscious and familiar to the human musical vocabulary "Vocalise." Biophilia as spontaneous feeling is the most dominant form of pantheistic behavior . . .

LASZLO: And the healthiest.

TOBIAS: Absolutely.

LASZLO: Professor Haffelder made a CD for me which includes all the elements of the sounds he played for me on the experiment, including music, but including also the whale call to which I had reacted so explosively. This kind of sound, he said, has the effect of tuning together our cerebral functions in such a way that we become more coherent, more sensitive, and altogether healthier. Every once in a while I put on earphones and listen to this CD. Stereo-earphones are important because it is not the same when you listen with the left or the right ear alone. You have to hear how the sound is actually transferred from one ear to the other. The effect of that is not something we feel immediately, but after a while we begin to feel "more together." The experience of listening to the call of the mother whale is a practical example of the importance of our relations to other species, and to nature in general. These sounds enhance and deepen our own wholeness.

TOBIAS: Yes. The response to Picasso's amazing painting *Guernica* with the rearing horse, which took him between nine and eleven days to paint, has been universally proclaimed as a riveting work. It elicits powerful reactions, probably because it embodies that chime of universal recognition. In 2011 the National Institute for the Clinical Application of Behavioral Medicine (NICABM) sponsored a conversation between two scientists interested in the neuroscience of psychotherapy, Dr. Louis Cozolino and Dr. Ruth Buczynski. The outcome of their conversation was a clear indication that the brain is a "social organ." They pointed out, for example, how taxi drivers in London have a larger hippocampus region in the medial temporal area of the brain because that primordial organ in the brain is known—as in other species—to replicate a map of the environment, and the more you rely on it, the larger the hippocampus. It must be enormous in squirrels, who remember a vast network of hidden acorns and the like. Imagine, then, the impact on the many regions of a human brain upon seeing Picasso's masterpiece: that silent, sprawling depiction of horror and grief. Art, like all other experience, concentrates and enlists our brains and our minds. Psychology and every form of therapy taps into such conditionals: the oldest stimulus/response models.

Another painting receiving such a response was the *Adoration of the Mystic Lamb*, the center panel of the unbelievable *Ghent Altarpiece* by Jan and Hubert van Eyck. It is arguably one of the greatest works of art ever pictorially represented, there in that humble Belgian cathedral. And again, in looking at that lamb, which is at the very epicenter of Christianity and also the core of all human spirituality, everyone who views that altarpiece is transported.

And just to go back a moment to the Jain jiva, that living soul in every microscopic dust mote, dew drop, grain of sand, or pre-Socratic atom. In the sixth century before the Common Era, you had a man named Mahavira, who spent most of his life walking naked from village to village, sitting down at each juncture in order to engage in conversation with locals about the merits of nonviolence, on topics relating to ahimsa and the worship of all living beings. This is a remarkable testimony to how ethical, activist visions can change our perceptions. In the *Acharanga Sutra*, which I mentioned earlier—one of many other Angas or copious works of literature flowering throughout Jain literary and philosophical history—we know of literally millions of pages of literature and science and ethics written by soul-searching botanists— the Jain monks and nuns—who psychoanalyzed aspects of the world down to the equivalent of the subatomic particle, and attributed specific characteristics to each and every mote of light and grain of sand. They would look into filtered sunlight and see exquisite life-forms.

This is, in a sense, a key to unearthing epigenetics and the ability of individuals to reshape their worldviews.

LASZLO: An interesting and meaningful vision.

TOBIAS: And it comports with the famous twentieth century letter that was sent to Mahatma Gandhi regarding whether he would kill a snake or let that snake kill him. And he said he would never kill the snake. I am simplifying it, but that was the question. Gandhi would never strike the cobra. And this was during the Ahmedabad period, starting in 1917, when he was really developing his political ideals. Today, ironically, Gandhi is almost forgotten by the youth of India. His legacy has been obscured by modernity in a way that is rapacious and in keeping with the same onrush of economic modalities and mantras that we see nauseatingly replicated from culture to culture in our own time. You have editorializing about how we have to somehow get the young generation engaged with Gandhi's message because it speaks to this time of homogenized and dumbed-down ethics more than ever. But this is why I ask you about biophilia, because we really are at a bifurcation point; we really are going into two simultaneous directions.

LASZLO: The ancient wisdom I am often repeating is that the future is not to be predicted, but created. We can do something about that—about creating how the world will truly be.

TOBIAS: Let me ask you about something in your book *What Is Reality: The New Map of Cosmos and Consciousness*. Stanislav Grof references those cosmologically embedded holotropic characterizations in his introduction to that work. In describing some powerful Buddhist revelations, he wrote, "The essence of the universe is succinctly captured in four statements. One in one, One in Many, Many in Many, and Many in One." I found this extremely powerful because it summarizes, in those four statements, everything. It says it all, in terms of our future. And it indicates what our veritable marching orders are, in some sense. How do you view it?

LASZLO: The proposition of one in many is not resolvable by mechanistic, fragmented thinking. The one is not different from the many, nor is the many from the one. They are different manifestations of a larger encompassing unit, the oneness unit, of which elements appear as a partial one, as a separate one, but these ones are not actually separate. The emerging concept is that we can conceive of reality as a manifestation of a holographic oneness where the information is distributed, so that in the information that constitutes of each thing includes the information that constitutes all things. This seemed to be just metaphysics before the discovery of the hologram in the 1960s by Dennis Gabor. The holographic principle tells us that we can have distributed information in real-world systems—that each and every thing includes in its essence all other things.

DAY FOUR

Afternoon, in the Study

LASZLO: About the holographic principle: it seems like an abstract, purely theoretic construct. But we can actually experience what it tells us: it is reality. Try this experiment. Take an old-fashioned slide projector and introduce a slide into it. Then remove the lens and project the image you get on a screen. You will have all of the elements of that image distributed over the entire area of the screen. This is the projection of the optical array. If you take the lens you had removed and place it anywhere over the screen, you will have the full image appear on the screen behind that point. The full information for the image is present in the optical array at all its points. The oneness is there in all the parts. You cannot separate the one part and the whole image. The whole and the one are the same at different scales.

TOBIAS: There are interesting iterations of that holistic divisibility, which also conjures the notion of fractals. I remember first reading the *Parmenides*, in which the one *and* the many, not the one *in* the many, became a guiding principle for early Greek philosophy. It is a fundamental dualism, non-divisible, that lends itself to any number of philosophical interpretations. In terms of approaching the river, it goes to the heart of the Siddhartha legacy. But the *Parmenides* to me was seminal, and what is transformational here is the "in," it's not the "and." The dualism offers no easy or logical alternative, notwithstanding the human quest for coherence, unity, and homecoming. Of course, Buddhistic tradition, for example, centers upon that very whole or full image you refer to. Other cultures and traditions tear it apart and seek out diversity, not wholeness.

LASZLO: That is right.

TOBIAS: You sparked a reminiscence in your book that fascinated me. Something I realized I forgot to ask you about—the difference between the Fourier transform and the Gabor transform. There was an extraordinarily enlightening portion of your book in terms of that . . . it's a different dimensional shift, totally, and I wish I had the text in front of me with that particular description. It was at the risk of simplification . . .

Laszlo: It's really a shift from the frequency domain to the spatial-temporal domain. The frequency domain is where you have individual wave patterns and you are transforming them into signals that you can locate in special-temporal coordinates. In a Gabor transform you don't have individual wave trains; the wave trains are integrated in such a way that they are fractally reproduced, one in the other, so the entire thing is one. The Gabor transform is a transformation from a hologram to an object, and in its reverse, from an object into a hologram.

Tobias: Another point: I call this the "Romeo and Juliet" codex. Not just Shakespeare's choice of locations for his great drama, but that legendary wall one can visit in the heart of the city of Verona. From the very beginning of your meditation on the cosmos in your new "Map of Reality" you wrote, "The vibrations of the bits and clusters of the basic reality of the Universe, they constitute the phenomena that meets our eye; embracing the basic clusters we know as particles and the complex clusters we know as organisms." Again, I revert to that human penchant known in the Jain pantheon as the "*samavasarana*," meaning a gathering together of like-minded species sitting in a most compatible group; the sitting together of the animals whose commensalist, symbiotic, mutualistic, and co-sympathetic orientations enable them to speak to each other from the heart and to act with unconditional love. Think of it as intimate communion at long last, and once again, as from the beginning of the human journey when we numbered only a few thousand across the planet, we got it right. We know that from Chauvet, Lascaux, and Shanidar 4, the cave in the Zagros Mountains in what today is Kurdistan.

We also know it from all of the parables surrounding Saint Francis and from that wondrous, aforementioned wall of Romeo and

Juliet, not that far from here as the roseate spoonbill flies—some 370 kilometers north by northeast—in Verona. The wall is covered in Jackson Pollack-like graffiti in thousands of doodles made by lovers throughout history. Their collective writings comprise one of the most compact concentrations of love letters and appeals from the heart ever written on stone. And it speaks to your clusters and to the relational conditions which you enshrine in your book as I have never read before in terms of the precise choreography between those bits before bits, and Romeo and Juliet.

Verona is one of my favorite little cities with a great human heart, one of the greatest. I love that wall. But I see in it this "chaos-absent entropy"—a mob scene that hosts tranquility, madness absolved of its madness. In fact, ultimately it is the Lascaux of post-Renaissance humanity, a depiction of everything one wants, hopes for, imagines. A vision quest, an odyssey, a hallucination. An encyclopedic rendering of the human story. Your approach to physics and cosmology seems to have helped us—all those like-minded lovers who have left their anonymous traces on that wall—to connect the words, graphics, and dots in a way that I think is very important. And I wonder how you feel with respect to the word *love*, coming from the so-called cold vacuum of space. Perhaps I am trying to load the question in favor of humanity, so to speak. In favor of such things as biophilia and love and our ability as a species and as individuals to express something that may be unique in the universe but is of the universe.

LASZLO: There is no such thing as a cold vacuum because there is no such thing as cold, even if the temperatures are measured and if, on the Kelvin scale, they are close to zero. In reality there is no such thing as a vacuum either. What there is, is a plenum, an energy plenum, an informed energy plenum. It is the information in the plenum that creates structure and coherence, and the experience that structure and coherence produces in and for us. The universe itself is neither cold nor a vacuum; it is a seething, turbulent plenum. But it is not chaotic because it is formed, or more exactly, "in-formed." It is a wholeness created by its "in-formation."

TOBIAS: Do you see that creation, that structuring, as a goal of evolution?

LASZLO: Yes, I do. The creation of coherent structures is, to my mind, the ultimate goal. Of evolution, and of existence itself.

TOBIAS: On the pages of your new manuscript I sensed a kind of structural epiphany whose underpinnings I would perceive as an entirely new interdisciplinary science of consciousness. You briefly discussed Gerald Edelman's book, *The Remembered Present: Biological Theory of Consciousness.* And you wrote, and permit me to quote extensively, what I thought of as a seminal section: "If consciousness is nothing but the result of brain functions, all brain functions can be assumed to produce consciousness. Does this hold even for relatively simple or primitive brain functions? Where do we draw the line? Chimpanzees and other primates behave as if they had consciousness, something that took zoologists and other life scientists decades to admit. . . . But do mice have consciousness, do insects have it? That is not clear. Plants do not have anything like a brain, therefore it is clear that they do not have consciousness, nor do primitive organisms, yet complex organisms are said to have evolved from primitive lineages and consciousness must thus have emerged from non-conscious foundations in the course of time. How this emergence would have taken place is not explained by the mainstream theory," unquote.

So this to me, Ervin, invites contemplation of that pivotal cornucopia of life within the Anthropocene, otherwise known as the "sixth extinction spasm" in the annals of biology. And it's the worst one in hundreds of millions of years because it is strictly determined by what I would term IPAC, an equation that is a version of the Ehrlich/Holdren IPAT equation devised in the mid-1970s, which declared that human impact on the global environment is defined as impact that is equivalent to population size, compounded by affluence multiplied by the levels of its technological punch. I've altered it to suggest IPAC, namely, that our inordinate impact on the biosphere can be most graphically assessed at the intersections of our population size, namely 7.4 billion human natures times the mean average of our inability to restrain aggression times that vastly fluctuating variable known as human consciousness, or unconsciousness as the case may be. And all within the scientifically understood umbrella condition of the late Garret Hardin's 1968 essay in the journal *Science* called "The Tragedy of the Commons," the realization that we are already consuming resources equivalent to four or five planet Earths, even at current population/consumption levels.

Now, you conclude your brief reflection of Edelman's book by stating, and I quote, "The mainstream theory faces an even more basic problem along these lines. This is the 'hard problem of consciousness theorists. How consciousness can be produced from something that is itself not conscious. Regardless of at what point in biological evolution consciousness would emerge in the brain when that point has been reached something as immaterial as consciousness arises from something as unconscious as matter. How could unconscious matter produce material consciousness?" unquote. So then this is very much at the heart of your astonishing linkages of such universal pillars as Planck's constant, the unblinking, unassailable, unshakable truths surrounding the brilliant time-bomb of a concept I think you have enshrined.

And I think it is a Nobel Prize-worthy revelation, namely, coherence, from the heart of the Big Bang to your very own heart, Ervin; from the unimaginably vast reaches of the cosmos to the vivid emergence of a biosphere whose every nuance, variety, purpose, and vibration achieves that same coherence in the name of health, vibrancy, and the prospects for a harmonious future here on Earth's remaining four billion or so years of life. It was fascinating to me how you did that. But it leaves me with a dangling modifier, obviously. A huge dangling modifier. I mean you've posed the question, but you haven't posed the answer, to my mind.

LASZLO: Trying to suggest the answer is to suggest a new metaphysics. We can adopt among others a metaphysics in which evolution is radical emergence. I, myself, don't believe that evolution is a radical emergence, it's not something entirely new coming to be, something that simply wasn't there. Rather, I believe that what is coming to be is an articulation, an expression of what there was. And what there was, and still is, is not space and time existing separately. What there was and is, is information acting on energy. It is the work of logos, orientation, spirit, intelligence—whatever we may call it. This something is beyond spacetime. It is itself unchanging, eternal. That's why I decided to re-title my book—first I called it "Beyond Spacetime." The key to understanding what was and is, is not a series of separate events happening in space and time. The higher or deeper reality is the realm Plato called the realm of the soul and the Hindus called Akasha. This realm is the basis of everything that happens in space and time.

Reality consists of the interplay of two realms or dimensions. In a previous book I called one the A-dimension, the Akashic Dimension. The other is the M-Dimension, the Manifest Dimension. I haven't introduced this term in our conversation yet, but I should note that in my view the best way we can conceptualize the nature of reality is in terms of an eternal, unchanging logos at one level, manifested through progressive change at another. This is a Platonic, but also a Whiteheadian, concept. Remember what Whitehead said: the history of philosophy is a series of footnotes to Plato. I didn't understand this years ago, but now I'm beginning to understand and I agree with it. As Plato said, there is a higher, deeper reality of which the perceived reality is a reflection. The deeper reality is the logos of the perceived reality.

There is no radical emergence in reality because the logos predates all things that exist, and that come to be. The perceived reality evolves from the deeper reality nonlinearly, crisis after crisis. It rises to higher and higher levels. What does this mean? To where would it lead?

TOBIAS: Hopefully not to another war requiem by Shostakovich, where we go from what I call the ecology of cemeteries to cemeteries. Sieges of Leningrad. Of course, my only problem with Plato is that he banished Aristotle from Athens and moved his school to what is today part of modern Turkey, while Aristotle moved back up north to his family's home on the peninsula of Chalcidice where he died alone, friendless, probably of some kind of colon or pancreatic cancer at the age of sixty-two or so. And as I said: banished from Athens by his own mentor, Plato. It's a strange historical fact of their relationship after seven years of studying together.

LASZLO: There are fundamental differences in the ways in which we can understand each other. There is no compromise possible. This applies also to our own thinking. I was an Aristotelian practically all of my life, ever since I was seriously engaged in philosophical inquiry. I was an empiricist. Then I saw the limitations of empiricism. There is a limitation in the understanding of consciousness. We cannot understand what consciousness is unless we assume that there is something beyond it.

TOBIAS: Well it would be very interesting to see how the patterns in nature and in human behavior as part of nature are moving towards the next republic.

LASZLO: You are asking for a forecast or prediction. And I said that it is not possible.

TOBIAS: Conservation biologists are able at this point in time, given the tools we have refined, to essentially forecast when a species will go extinct. Pretty awful.

LASZLO: That prediction does not include the factor of human consciousness. In regard to the question of the future of conscious humanity, the best we can do is to see what the conditions are that would lead us with a significant probability towards what you call the next republic.

TOBIAS: We have stochastic targets and probability distribution. We have what I have termed in a new book of mine, "Ideal Algebra," that enables us to speculate with some precision.

LASZLO: We can move backwards from a target that we want to achieve and see the conditions that enhance the probability of achieving it. This is the best we can do.

TOBIAS: I believe in paradise. I think we can get there. It's incipient.

LASZLO: Yes, but will it come about on Earth?

TOBIAS: Ahh! Well not in these parts of the world, that's for sure. Not yet. And certainly not in the former Bangladesh that I experienced and suffered firsthand.

LASZLO: At the same time this marvelous person, Muhammad Yunus, is from Bangladesh. He initiated the policy of granting microcredits, something truly positive and important. I had the good fortune to meet him.

TOBIAS: I had a wonderful coincidence one day after flying home from Bangladesh. I was invited to a dinner party in Malibu, of all places, where lo and behold the guest of honor was none other than Mohammad Yunus. So, there I was having dinner with him.

LASZLO: Let's take a fresh breath and move to another level. I have expressed many times what I consider fundamental ideas I had in my head. I have said those things and I am not aware of any other major ideas in my head at this point. But I am happy to hear what you have in mind and go further.

TOBIAS: I think the major premise that excites me and challenges my thinking is really the dialogue between physics and ecology. Not so much in the energy sector, however, because those issues are obvious—

LASZLO: People discuss them all the time.

TOBIAS: Yes. And this is basically the sustainability mantras that are so prolix that it's almost glib now. Yet it's a realm of technology innovation that is not central at all to where I'd like to see this to go. See how this relates to the universe. It's that basic sense that first perplexed me with regard to your work in terms of speaking of the universe, of subatomic particles, of cosmology, and the concern I have from an ecological perspective, which to me is (with respect to our crisis on this planet) THE crisis. It is the matrix for all of the lamentations and jeremiads and real suffering that can be quantified or qualified in every domain of human experience (Dürer spent his entire life, in a sense, painting the Anthropocene through the Passion of Christ). It is the anthropic principle, which for so many years and centuries and millennia has been at the forefront of human thought because we think so highly of ourselves.

Take but one of thousands of examples—Respighi's lovely little musical sketch, "The Nightingale" from *The Birds* suite, for instance. That lovely piece of music is analogous to the ego and superego in breaking down the barriers to unifying ourselves with others, including Respighi's birds. It seems to me that first of all we have to let go, and in letting go we have to truly let go, period. By doing that we're breaking down barriers right away. Barriers meaning the gulf between us and the other that I had referenced earlier today. You've talked about it, for example, when you mentioned the acoustical sensation you received of hearing the mother whale crying out to her young. To me this is the biggest, the most urgent peril we face: our inability to reconcile ourselves with the other. And you speak, for example, of Planck units, Planck scale, and these exponentially minute edges of the universe, both in time, space, size, ratio, and proportion (given the fact that the universe keeps expanding in volume), yet the ratios, the harmonious ratios apparently remain the same throughout those distances, those spaces, those enormities of time and space.

Such glorious harmonies are in our minds, as much as we want to be assured they are real. Although they may mean everything to physicists and cosmologists, I would wager that most people are not

particularly focused on them, certainly not from nine to five. I suspect that what means something to most people is a glass of cold,
clean drinking water. So how do we bridge that vast gap? People look
with wonder at the Hubble Telescope images, or ten years' worth of
images of the surface of Mars, and they can be amazed by astronomical discoveries. But if you remember Paul Zweig's book *The Adven-
turer*, he was the one who said that within days of the first landing
on the moon, people were already bored by the idea. The news came
and went. The second Apollo got less press, and the third even less.
So now NASA has created its own TV and radio channels to try and
sustain interest from the public, particularly in terms of galvanizing
Congress to perpetuate funding, with particular energy in areas of
NASA's spectacular climate change imaging across the globe, but also
a Mars mission, the mission close to the Sun, the study of Saturn's
magnificent and unimaginable rings, and so forth.

Of course, mainstream news gets its adrenalin from disasters, from the *Challenger* tragedy to private rocket ships that have
exploded in the last few years. Meanwhile, movies like *Gravity* and
The Martian aim to remind us how precious the earth is. But I'm trying to see how we can relate in the most gravid manner this vast array
of harmonies that you've been, both in detail and in general, system-
atically elaborating upon in your work for so many years, with the
present eco-crises. And to me that's what certainly demands a full
accounting, if possible. It needs much more elaboration, reconcilia-
tion, integration.

LASZLO: It certainly does.

TOBIAS: It is still begging for help in clarification and making more
pellucid, more tangible the relationship. And again, we spoke this

morning about the opportunity our species finds itself poised upon. It's an opportunity that is also a terrifying chasm, a cliff off which we could easily plunge.

LASZLO: We are moving through different aspects of the topic, whereas we should get to a point where we can address the basic problem.

TOBIAS: For me a basic problem is how we relate cosmology to environmental issues.

LASZLO: Cosmologist Steven Weinberg said more than ten years ago that he is convinced that the more we learn about the nature of physical laws, the more we learn that this is a cold, impersonal universe that has no place for mind and life. The universe is just a series of physical events that ultimately comes to an end, for the universe runs down. That was *the* dominant paradigm until the current revolution that is discovering nonlocality, entanglement, and non-perceivable wormholes among physical events, as well as non-evident virtual energies. Ten years ago still, the dominant concept was the universe as a giant system that runs down. There is no place for life in such a universe. Life must be an accident. It could happen in some places through a series of fortunate coincidences, but very rarely. That was basically the problem that Ludwig Boltzmann grappled with at the end of the nineteenth century. And it was the problem that occupied the minds of many leading scientists and philosophers at (and also beyond) the turn of the twentieth century—Henri Bergson, for example.

TOBIAS: Bergson's "élan vital."

LASZLO: Yes indeed. But the new paradigm fundamentally changes our concept of the physical universe and overcomes the problem of the dichotomy between physical process and life.

Tobias: Did Carl Sagan have a role in at least popularizing the bridge between life here on Earth and the prospect of life elsewhere? Did SETI, the Search for Extraterrestrial Intelligence? Does anything for you access here the new ideas?

Laszlo: Sagan helped to create a sense of wonder about the universe and the realization that we can't be all that exceptional. There must be life also elsewhere in the universe.

Tobias: Well exceptionalism is definitely a key word in terms of the "anthropic principle" and the whole characterization of ourselves and our life on this planet as unique. That has been, wouldn't you say, the dominant mindset?

Laszlo: "Unique" is even stronger than "exceptional." We can be exceptional in some ways but not unique—if we are the manifestation of something that is universal and is seeking expression throughout nature. This is the key to how we now look at the cosmos. What is the basic nature of reality? If the universe is a giant machine that runs down, then we, and all life in space and time, are freaks. If physical reality is a manifestation of a cosmic intelligence that seeks to actualize itself, then we are exceptional only in being a high-level manifestation of that intelligence. This conception, however, was banished from science. It was said to be metaphysics.

Tobias: Hoddinott's *Folksong Suite* is a love song likely known by very few. I heard it I think for the first time just before coming to your home here in wondrous Tuscany, just the other day, performed by the Royal Ballet Sinfonia with Andrew Penny conducting—so quiet, gentle, and purely of a human dimension. I can well see it contradistinguished against a paradigm in which the rest of the universe is deemed to be alien and heartless, without a human signature even remotely on the least horizon. As beautiful as the night sky is, as warm with constellations, a human brain seeking emotional connection is not going to possess the mirror neurons that would enable one to cozy up to a star the way we might do with, say, *The Death and Assumption of the Virgin* or other paintings by Fra Angelico. This is where it behooves the biggest ideas and ideals of science to address the nature of human intimacy—which is emblematic of all intimates throughout the zoological worlds with which we cohabit.

LASZLO: The latest advances in science demonstrate a level of coherence that suggests wholeness—a wholeness that not only *is* in the universe, but that evolves. This is very different from the idea that the universe breaks down and moves toward entropy. This universe is a quasi-living entity. Duane Elgin among others argued this clearly and beautifully. But the living universe concept is not accepted and integrated in science, it still seems like a poetical, metaphysical notion. Yet the basic concept is gaining acceptance—the concept of wholeness in and among phenomena. That wholeness hallmarks the nature of all life.

Think about this for a moment: Is this not amazing, coming from physics? It suggests that integration, unity, and harmony are natural phenomena and that they may culminate in the emergence of life. These phenomena come to expression already in the evolution of stars where, inconceivably, organic molecules prove to be present and chains of carbon link up to produce the foundations of living systems. Wherever there is an opportunity to create living systems, evolution seizes on it. Life, it appears, is part of the cosmic plan, part of the self-actualization of the universe.

The universe discloses a design, whether or not we attribute it to a supernatural "being." We can, of course, ask ourselves who or what is behind the laws of nature that are so incredibly precise that a change even of one billionth of a part of the relationship between, say, the mass of the proton and the expansion of the universe would change the evolution of the universe. It would render impossible the evolution of stars and therefore the energy flow from hot stars to planets, the flow that is basic to life as we know it. This is a very coherent universe. Can its coherence be accidental?

TOBIAS: What is the implication of that? If it's not an accident, where does that take us?

LASZLO: It takes us to a conception that life is a manifestation of something basic to the universe—a cosmic intelligence. Nothing is here by chance.

TOBIAS: But again, the big question: Where does that lead our species? We are the youngest large vertebrate species on this planet, and at the rate we're going, we are in danger of having the shortest-lived epitaph in the annals of biology. So this collision course that we have put ourselves on, in a place that you have described as not an accident, is positioned in such a way as to advance the cause of self-expression at the highest level, and is one that mirrors and represents perhaps a cosmic unification that is not miraculous but perfect. And yet, when I think of it and I remember Mandelbrot—I had long discussions with him—and I remember saying once to him that I thought the fractals were beautiful but totally sterile and he said well, yes, it is a replication process, and we don't understand it fully, and yes, for him it was a mathematical art form. I think it is fair to say in retrospect that this is a turning point for humanity. If we gain some kind of relief or draw inspiration from this cosmic relationship, we're not pen pals, we're caught in a tight web . . .

LASZLO: What do you see when you look at a human being? As an ecologist, you probably look at the living organism.

TOBIAS: Yes.

LASZLO: The organism is a set of highly complex relationships inserted into an environment where it has to maintain itself. If you look at a human being in that perspective, you are looking at one side of the coin: the physical expression of a cosmic trend. This is surely fundamental, but it is only one side.

TOBIAS: What would be the other side?

LASZLO: The mental. The ideational, the "Idea" as Plato called it, with a capital "I."

TOBIAS: Are you now speaking as a philosopher or as a scientist?

LASZLO: I am trying to speak as a new paradigm scientist. The new physics tells us that the laws of nature are algorithms coded into the information that pervades space and time, and if we say that this information is mirrored, represented, expressed in our consciousness, then we are speaking as the great scientists sought to speak. Einstein said that to read the laws of nature is to read the mind of God. If we are looking at the body, we will find that how it's evolving is toward complexity. Sets or clusters of vibrating energies come together to create sets and clusters of higher and higher coherence. But if we are looking at evolution as a process of emerging coherence in both body and mind, then we see that evolution is also the evolution of consciousness.

In the holistic perspective what is happening in evolution is something we can't describe in physical terms because there is no way we can understand our stream of sensations as produced by the interaction of neurons. If we are looking at both sides, we recognize the intimate relationship between mind and matter. There is something other than the purely physical emerging in the world. Consciousness is evolving from the proto-mind Alfred North Whitehead described as a "prehension" that is present in every "actual entity." The concept of a proto-consciousness evolving in the world is present in the philosophy of Spinoza as well, and in all the great Hindu and Buddhist systems.

TOBIAS: In those systems the old adage that the origins of consciousness is pain has always been fundamental. Buddha said it. It is the basis of Jainism. It is the most mortal and potentially unrecoverable of concepts. It was an idea whose time has come clearly for our species. But my question is, if we are in a transformational point of departure at the very moment that we have this profound opportunity granted us through the lenses of science and the technology and the reflection that has been prompted by these discoveries, do you feel there is, in fact, some pragmatic, remedial, therapeutic acceptance at long last by our species?

LASZLO: I have serious doubts whether we can transmit it and open the path to a transformation in a painless way, to use the term you just used. We have gone too far now to make the transition painless. There are going to be species transformations that include extinctions, but could also include regeneration and rebirth. In any case, it will not be a simple linear unfolding of an evolutionary trend.

DAY FIVE

Morning, at the Upstairs Terrace

TOBIAS: We were speaking of a global transformation. I think the question is not whether, but when. Do you envision a time frame in reference to where we are now?

LASZLO: At the most a few decades, but perhaps much less. It could be a matter of years. I can't imagine that today's human system will become sustainable without major shocks leading to major changes in consciousness.

TOBIAS: Would it be injurious to the universe if our species goes extinct in thirty or forty years?

LASZLO: If the universe is a progressive articulation of mind or consciousness, it is a branch of the evolving consciousness that was highly promising not long ago. It would be a pity if our species were to die out. It would be injurious to the universe in that the trend towards self-actualization, toward evolutionary unfolding, would be set back.

TOBIAS: And that would be the only branch?

LASZLO: I don't think that we are of such cosmic importance that we couldn't be substituted by life and intelligence on other planets. Yet it would be better not to test this by dying out. We should learn from our mistakes. If we keep making mistakes, we become extinct. Natural selection works on us, as it works on everything. On a simpler level of evolution there are no consciously committed mistakes. But when a species reaches the level of consciousness where it can entertain alternatives, it can make mistakes.

TOBIAS: Correct.

LASZLO: An evolved species is capable of adopting behaviors that are detrimental to its survival, as well as to the existence and survival of other species around it.

TOBIAS: Yes, we have that power and we are exhibiting it.

LASZLO: We are making mistakes due to the power of our consciousness to choose among alternatives, Unfortunately, we have been choosing some of the wrong alternatives. That is a threat to our survival.

TOBIAS: Yes. If that is the case, if that time line is so tight, with which I happen to concur, I mean I don't consider that pessimism; I consider that ultrarealism. I think that the entropic nature that we demonstrate at every level of our paradoxical fallacies, as I liken it, the situation we find ourselves in is so vulnerable and fragile right now. I could not have imagined even ten years ago this level of desperate peril in which our species and the entire biosphere is in. We have solely perpetrated this situation. If our time frame is so limited that we have not a hundred years, or even fifty years left (and many, if not most, biologists would probably agree with this prognosis), if there is a large biospheric implosion just around the corner (and we are seeing glimpses of it every minute), then we have to move quickly and re-envision an alternative, as you used the word. And I don't mean the mundane alternative energy—we need a fundamental shift in human nature, what my mind cries out for, a *new nature*, and one that we need to give birth to rapidly, given all of the different indices that are discordant, that are providing little evidence of positive congruence.

Indeed, coherence seems to be breaking down in every human or human-impacted sphere. We are sorely reminded of that incongruence and incoherence every day. Our multiples of depression, a worldwide epidemic of depression, stems in large part from the irony of our helplessness to effect the change we wish to be. I say irony because we are the agents of the investigable causes of that dystopian world we lament, to use a concept from a fine book by Calvin O. Schrag, *Radical Reflection and the Origin of the Human Sciences* (1980). In critiquing Habermas and Husserl, Schrag writes, "Habermas finds all science, both that of nature and of man, subject to the illusion of pure theory, cut off from its roots in the domain of human interest, and thus placed in a position of being unable to understand itself."

Curiously, we think we understand, at least partly so, the roots of our depression as they bear directly on the full weight of the global ecological crisis and our dualistic relationship to nature. But we are mostly in the dark when it comes to the devil in the details. We can rattle off certain statistics—the amount of rainforest being destroyed by humans annually, for example. But approximately 97 percent of the human population is almost entirely out of touch with the intimate realities of living in an even partially pure state of nature, unlike those nearly 300 million indigenous peoples remaining on the planet.

If I go to a university campus and ask any classroom, "How much time do you all feel we have left as a species on this earth?"

they think I am half crazy to ask such a question. This is especially so in the campus environment, where one still imagines and breathes in the aura of the Sorbonne circa 1200 where everything is possible, where a college education is the ultimate liberation of ideas and of ideals, and where life is blooming with possibilities. Not so in today's ecological reality. But if I say that or in any way promulgate a dark and dismal vision, I get booed off the stage. Or, conversely, an auditorium of very depressed students utters not a single word; they all agree with me. All or nothing scenarios, in other words.

If I had to summarize my experience with college students, I would say that for the most part they seem to feel empowered. Is that positive, or is that a blind spot? I don't know. I don't have the answer to that, not off my hip. But from a sociological purview I am aware that many of my peers and the younger generations seem to be oblivious or indifferent to the blank check condition of human existence. I attribute that to an escalating series of mood swings everywhere discernible, from the classroom to the Congress to public polls.

LASZLO: Just look at the dynamics of evolution on the biological level. Population biologists such as Niles Eldredge and Jay Gould argued that the dynamics of human populations are the dynamics of complex systems in general. It's not the Darwinian dynamics based on chance, and it does not produce a linear transformation. The dominant species in a biological population does not change—it just dies out. If that species makes mistakes consistently and is not replaced, the entire population of an ecosystem will become extinct. The alternative is that the mutations that are constantly occurring at the periphery are empowered to enter the space left by the outgoing dominant species.

TOBIAS: They will occupy the dominant niche.

LASZLO: That is the nature of the systems dynamics of population evolution. It applies to the evolution of the human species. Here, too, we cannot expect that the leading elements will change and transform all by themselves. What is likely to happen is that the level of crisis becomes so high that the system itself is imploding, and those who are leading it are no longer effective because their power to lead has been impaired and then removed. In the end, they have nothing to lead. At the same time, there is constant evolution at the periphery, and these are not necessarily biological mutations; they are cultural mutations. They are mutations of consciousness. Changes in the mentality that defines how people behave. They are decisive elements in an unstable system where the behavior even of a small group of people can modify the evolution of the whole system.

TOBIAS: But that seems contradictory to current experience. For example, you referenced leadership. Political leadership is typically three, four, five, or six years. If we are talking twenty, thirty, forty years as our maximum horizon for survival, then we are talking three or four more parties in power. Ever. Speaking literally now, politically, we're talking about a single generation in human terms. That is fifteen to twenty-five years given the reproductive proclivities of our species, where puberty is reached sometime after eleven years of age. For certain ethnic groups of Central Africa there are cases of menarche and some level of reproductive success occurring as early as nine years of age, but worldwide we're usually talking fifteen to twenty-five years, okay?

And we know, for a fact, that between now and one generation from now we're going to add two billion more people to the planet, two billion more high consumerist individuals that want nothing more than you and I want. Western style aspirants who aren't going to be the ones who have any power hold or any manipulative ability to reach geographically beyond their immediate household or, at best, neighborhood. They may have great dreams and ideas, but practically speaking if we have so little time, and I think we do (and I'm not alone in that persuasive ideology, based on voluminous and good data sets), then we're looking at a dead end unless there is some radical transformational power that lures us in and persuades us with some form of sublime alternatives that we can hang our hats on and believe in.

LASZLO: The power doesn't come from the outside. The power of transformation comes from inside and is triggered by crisis.

TOBIAS: We are in the midst of that crisis. We know that.

LASZLO: Yes. But the decisive factor is the level of the crisis. It does not have to be devastating. A non-acute, non-life-threatening crisis can also trigger changes already because people are becoming aware of it. "Perceived crises." Such crises need not be fatal, or even violent.

TOBIAS: We know that we're in the thick of the Anthropocene. There is no question about that. The rate of extinctions is unprecedented. And we are the sole cause.

LASZLO: We are talking about an already perceiving crisis, and we suspect that it is triggered by obsolete leadership. This calls for triggering systemic change—a transformation. The people who are in power are seldom open to transformation. This is the dynamics of change and evolution also in a population. The dominant species just maintains itself. Until it collapses. Or becomes extinct. But new cultures are growing up, new generations are springing up at the periphery from the ground. This doesn't necessarily mean young people—the new cultures can be people of any age. But they are more likely to consist mainly of the young, because they are the ones who do not have a vested interest in maintaining the dominant culture.

TOBIAS: Ervin, we've just laid out the numbers here. The time frame and the constraints. You spoke of the time frame and the constraints decisively in terms of why life is not an accident, referring to the 13.8 billion light years and the complexities of trillions of electromagnetic events, of the very protons, nuclei, the whole atomic chain of events leading to (among a myriad of other destinations) molecular biology, consciousness, and the conscience. It couldn't be an accident. We are now speaking about a dominant, carnivorous, predatory species—humans—which happens to be, for the time being, unambiguously the dominant species, with the exception of most pollinators, social insects, and spiders. Just last week in the journal *Science*, they published data from the last four years. Every predator on this planet over fifty kilograms that they have studied, which means the large predators, are going extinct or are on the verge of extinction because of us. Every one of us. We are the dominant ones.

This conquest-driven fallacy in ourselves is why the great, late Aldo Leopold, describing the "community concept" in *A Sand County Almanac* (1966) wrote: "A land ethic changes the role of *Homo Sapiens* from conqueror of the land-community to plain member and citizen of it. It implies respect for his fellow-members, and also respect for the community as such." We seldom peer deeply inward like Leopold, except to say, "Yes, we are the fuel that is driving this Anthropocene." But there are very few who are willing to concede that we are at risk, because you don't get research grants for saying those things. It's very dangerous to one's continued funding. Nobody wants to hear that we are the problem. I know colleagues who are waiting until they retire to say such things. And because there is this basic knee-jerk reaction across all of the sciences and all of academia that is the leadership system that says you don't want to daunt or chill the upcoming students, the children—you want to give them hope. It's almost a blind mantra of hope.

LASZLO: But this hope is paradoxical. Real hope is in the crisis.

TOBIAS: From a very wise vantage point.

LASZLO L: You have to recognize the dynamic. The dynamic is biological evolution, paralleled by the evolution of consciousness, and these evolutions encounter crisis points, nonlinear cusps, and these are the templates of transformation.

TOBIAS: If that's true—and let's suppose for a moment that we have a single collision course happening—it's not multivariant, and we know it to be the sum total of many variables that have gone negative, not positive. And the synchronistic summation of these variables, the population, affluence, consumption levels, and so on, are all pointing in a direction that can be summarized in a singular crisis, as a critical singularity of our species. We are at the juncture; we are theoretically at the end of the road, biologically speaking, at the very moment that we are expressing it all. It's a self-referential doomsday, which André Leroi-Gourhan explained in his 1964 book *Le Geste et la Parole* by reminding readers that the human community today, with regard to the war-making apparatus, is "hardly different from Homo sapiens of three hundred centuries ago."

This is the paradox, in my mind. We are fully articulate regarding the minutia of our war-making against the planet, this Anthropocene epoch we are in, even whilst accelerating its worst particulars, of which we are the sole agent. We're sitting in a comfortable setting on leisurely sofas with all the preferred trappings and ingredients and components of human existence, here in Tuscany. And this dialogue, as we both know, is the heir to trial and efforts gone awry and something like ten thousand generations of our ancestors behind us enabling us to achieve this, and here we are recognizing the bifurcation before us and speaking of the younger generation, not necessarily younger in years, but certainly in terms of enthusiasm and willingness to change, willingness to embrace new ideals.

LASZLO: Indeed. The young generations in most parts of the world manifest openness.

TOBIAS: Yes. So, we're at this incredible turning point, and if we should fail to make that change—this gesture of "manifest openness," as you call it—then we are royally at fault, and future generations will look back in awe at our irresponsibility. Remember, those future generations are fast upon us and already looking back at us from the near future. This turning point is a prime conditional that can segue into pain or joy. There are 7.4 billion individuals, 7.4 billion human natures; you might argue that we're all genetically basically identical, yet a single decision by a single individual can change everything. Napoleon can order hundreds of thousands of young men into Moscow in winter and see them never return. Although he is said to have quipped upon that very disaster that one night of rabble rousing back in Paris would, through the peak season of tens of thousands of procreations, restore the number to his insane army.

But there are singular decisions by individuals that have left horrifying marks, needless to say. One person giving the go-ahead to the Enola Gay, or making a sign that reads, "Work Sets You Free," and so forth. And even these grotesque, unprecedented decisions by people in power have not altered demonstrably human demography.

Most nations in sub-Saharan Africa, for example, are surging ahead in terms of their total fertility rates (TFRs). The vast tragedies of HIV and Ebola, or the legacies of slavery, liberation armies, and child slavery, endlessly into the night, have not changed the fact that countries like Ethiopia and Nigeria are adding massive numbers to a continent already churning under severe environmental constraints with respect to sustainable resource development. In most of the trouble spots, most of the nations that are nondemocratic are engulfed in poverty and ecological turmoil, and their populations are increasing, not diminishing. Their TFRs are still well over three. In a nation like Niger, the TFR is still 7.6. So we see crisis points; there is no doubt about it, and these cumulative crises are resulting from individual human decisions. Every man and woman who decides to attempt to bring another child into this world are making a choice, and of those newborns 80 percent are destined to be born into poverty with the resulting added burden on habitat, wildlife, fresh water, and so on.

So those are just some of the indications. We just saw the latest Population Reference Bureau and UN Population Fund reports. I follow the demography closely because it is a core ecological crisis. You can have all the best conservationist practices on Earth, but if the TFR continues as it is, fueling a runaway population train that is clearly going to reach ten billion, possibly even eleven billion, of ungainly, largely carnivorous *Homo sapiens* by the century's end—that is assuming we adopt an optimistic attitude and assert that there will be, for humans, a century's end—then all of that conservation is going to be undermined. All the hard work by those thousands of NGOs and humane societies and tree-planting organizations will be for naught. Yes, that is solid pessimism, but when you look at the IPAT equation by Ehrlich and Holdren that I referenced earlier, one can't help but put the pieces together and see clearly what's unfolding within this very generation. It's that empty stomach syndrome. You are not going to persuade somebody not to kill that last mountain gorilla or rhino if they are hungry, or greedy, or worried enough that they are willing to take the shot and earn a year's income with a single bullet. This is Hardin's "Tragedy of the Commons," his theory, as well as the famed 1968 essay he wrote by the same title for the journal *Science*, taken to its most logical ending, although it is really more like a desperate poem by T. S. Eliot. The time frame is what worries me, Ervin.

You've posited a relationship that is intrinsically magnificent, phenomenal, and shockingly original, and the synthesis that you've brought to bear upon it in a multidisciplinary and cogent contextualization, is methodically breathtaking. As a working anthrozoologist whose life is dedicated to ameliorating pain, helping other species, helping our species to help nature and to be inspired by the environment and join in the great challenge of this generation, which is to save what's left of miraculous creation, I am worried. And what worries me most and is the key to all of this and the reason I wanted us to have this dialogue, is that we must grapple with the complexities of the cosmos here on Earth, on this earth, not some posited otherworldly Earth that may in some mathematical equation exist according to some fragment of a probability theorem.

We are in the red zone, the hot zone. We've hit the flash point. And we're going down and taking every vertebrate with us. I exclude invertebrates because they show much more marked biological resiliency. Even the early Woods Hole research suggests—as does much more recent data from Chernobyl (that cockroaches, for example, are not necessarily shaken by high levels of radiations)—that various microbial life-forms can thrive in 170 degrees Fahrenheit surface temperatures on the sands of the Namib Desert, in thermal vents in hot springs in Yellowstone, and in Iceland and in deep marine sulfuric outgassings. Those bacteria and viruses and prions appear to have a level of immunity to our woes, but it may not be much solace to those of us who are vertebrates unless all our philosophies can turn into something interesting and more compelling, even, than our own survival (to paraphrase Shakespeare). So, I am not worried about evolution continuing for a few more billion years on Earth utilizing these microbial stewards as its champion. We couldn't extinguish evolution.

LASZLO: Not on Earth, and not anywhere in this universe.

TOBIAS: That's right. But what I am worried about are the Vermeers and Rembrandts and Tuscany. We should have learned by now. I mean we've had these magnificent lessons for thousands of years. The Australasian indigenous peoples have had the rich cultures of their Australian aboriginal forebears returned to them through new technologies introduced by French paleontologists. These technologies have made pellucid the heretofore nonvisible petroglyphs made thirty thousand years ago. It's quite exciting to contemplate the prospects for other cave painting revelations throughout the world previously deemed to have been only oral transmissions from elders. Now these stories can be viewed on rock surfaces as well. My point is, for tens of thousands of years we've been reading the book and it is a wise book. By "the book" I am metaphorically referring to all of the received inspirations of our collective ancestries and genealogies, the many anecdotes and caudal residuum of tales within every family that has suffered and lived and created and investigated and provided nurturance for the subsequent generation. We've passed down this remarkable story of humanity to one another, and yet, we still haven't learned. Ultimately, we are in far worse shape than ever before. Now, when we add to this legacy that of fire, the new fire power of thousands of nuclear weapons, along with the time constraint of a few decades, how do you see us eluding this cul-de-sac of extinction? I mean, Ervin, how do you see us making it through this infernal chute?

Laszlo: Surely not without a serious reduction of the population. But there could be vast emergent trends bringing about a rewiring, so to speak, of our relations with one another and with nature.

Tobias: Do you believe that?

Laszlo: The very poorest populations, the most underprivileged people, will be difficult to induce to change; and people at the top will be difficult to get to change as well. The poor cannot readily change; they have few alternatives—deprivation has reduced the available options. The rich, because they have a vested interest in the status quo. The poorest and the richest may not be able to master the art of changing in time. But there is a very large core that will be able to shift, and I believe will shift.

Tobias: What do you mean a large core? Can you translate that, approximately?

Laszlo: It can be anywhere from a few percent to 50 or even 60 percent of the world population.

TOBIAS: In twenty years 50 percent will mean nearly five billion people. As for 1 percent, well, do the math, it's pretty basic. And as I mentioned before, some seventy thousand years ago when the Toba supervolcano erupted we got down to some twenty to forty thousand people, but since then we've come back. And that was an unexpected geological crisis, of course, which engendered something like a two-year nuclear winter equivalency around the globe, plunging our species into this genetic squeeze. So we managed . . . it's actually quite amazing to think that ten to twenty thousand breeding pairs of *Homo sapiens* survived. And yet it's not amazing when you consider the ample resources at their disposal.

It was Thomas Jefferson and the Louisiana Purchase that specifically enabled a rapidly growing North America to suddenly acquire countless states and several Canadian provinces overnight for the meager, subpar, sub-blue book price, so to speak, of around sixteen million dollars. And this was all because of the first successful Caribbean revolt by people of color on Napoleon's former Hispaniola. It was the deal of the century—not quite constitutional—but Jefferson knew that the number of immigrants coming to the United States was vast. I realize I'm leaping from 70,000 years ago to 1803, and from what is today part of Sumatra to the United States, but my point is this resiliency, even at the core of human demographics and the human genome, in terms of our ability thus far to survive one way or another. And this resiliency is there, even as I read the rap sheet of our ecological predilections and woes and the Anthropocene extinction patterns that would seem to all but doom us, and the current business-as-usual paradigms we have, as a globalized culture, all but embraced.

And along comes a person named Trump, the shadow of Napoleon who, for his own twisted, tortured, probably psychotic reasons, figured he simply didn't need the Port of Louisiana and all the surrounding territories. We don't have anywhere to go anymore, and we know we're running out of drinking water and arable land. We're burning down and drying out the world's most significant genetic havens, from the Great Barrier Reef to the Amazon, at the equivalency of one Belgium per year, or more than 7,500 square kilometers annually. And that does not begin to account for the road-effects that expedite the massive shrinking of the second largest tropical forests in the world, those in the Congo.

These trends are not diminishing but spiking. Brazil is in a new recession, as is Russia—the former tied to overproduction of soy, the latter tied to oil and gas prices on the global marketplace. And all this not to mention the current devaluing currency in China. So all of these seemingly mundane passing economic ephemera are tied to geography, which in turn immediately declares more facets, moment by moment, of the ecological and demographic crises. I've been trying to get a handle on it with this broken crystal ball between bouts of huge depression and resurgent appeals from the heart to something akin to hope that we will make it in spite of ourselves.

Laszlo: You don't sound like you believe it yourself.

Tobias: All these cumulative particulars are not boding well. We are seeing the rise of a new fascism in Republican and Tea Party American politics, which is significant since America is the dominant economy on the planet. It's interesting to note, however, that the TFR in the US is not on a par with the demographically imploding nations of Europe or the Pacific Rim. The US Census Bureau has estimated that at current trends, the United States could indeed exceed 400 or even 500 million people by the end of the twenty-first century—again, assuming there will be an end of the twenty-first century for humans. And these numbers are not attributable to immigration, as much as many of the extremist cultural xenophobes in the United States would have you believe. The US still exceeds most other Western nations in teenage pregnancy rates as well as unwanted pregnancies, which account for around 50 percent of all pregnancies in America. It is still not politically correct to have decent sex education in schools. Additionally, *Roe v. Wade* may well be up for grabs in a Republican-dominated Supreme Court, with state after state embattled with Planned Parenthood, which could well be stripped of its funding. This all threatens 175 years of the struggle for women's rights.

We have a lot of issues here, and they are not welcoming with respect to this paradigm shift. So as an ecologist, this all worries me. I am not a politician but I try to stay on top of the basic headlines because the time constraints are shrinking. Scientists need to be vocal activists. Our biological lifelines are dissipating, and our hopes for the future are narrowing. And this is why I am searching personally, as I know you are, and have been throughout your entire career.

Clearly you have come from (I am surmising) a world of great beauty where you honored beauty with your own practitioner's love of the craft. And you have come from that and entered into this realm and found yourself through the years, as I perceive you, deeply embedded in the concern for the future of this planet, for the future of life, and for the future of humanity. You could not have performed music on a magnificent instrument if you didn't love art and what produces art, which is humanity and the evolutionary pulses that have chimed the word "humanity." So clearly the peril that we are describing and discussing affects us both personally, as well as our loved ones. I mean we're talking about our loved ones here, aren't we?

LASZLO: I think our thinking will change when we think of our existence on this planet in a perspective that goes beyond single lifetimes. An impact on human thinking and behavior motived by the recognition that "we do not just go around once" may seem far out. But we should remember that single-lifetime thinking is peculiar to modern Western civilization. Traditional people and people in non-Western cultures don't think in these terms. Nor do animals, insofar as we can speak of their "thinking in terms" of something. When an animal comes into the world it doesn't think it has come into a single, finite, linear existence, born out of nothing and falling back into nothing. An animal has in its subconscious mind all the experience its species has accumulated in its previous existences. There is continuity in the kingdom of life through many, very many, lifetimes.

This subconscious memory-store is a vast resource that is simply not accessible if we think that instinct is inexplicable, that veridical intuition is just fantasy. But our connection to one another and to our past is always present because our mind is a repository of the collective experience of our forefathers and of the entire human species. We possess the collective wisdom of humankind, and that is a fabulous resource.

It's not the single-lifetime individual who needs to change, but the individual who is able to draw on the collective resource accumulated in the span of untold generations. The past is present; it is influencing us. "Inborn" knowledge and spontaneous skills, familiar from the animal kingdom, are not inexplicable and mysterious. They are the accumulated memory of lived experience over many lifetimes. How capable are we to recall that tremendous wealth and draw on it? It is up to us. That resource, that potential wealth, that natural wisdom, is a crucial factor as we confront the life-deciding issues of today's world.

TOBIAS: Is this not something Heidegger acknowledged when he spoke about history, a very specific kind of lived experience?

LASZLO: The concept of accumulated knowledge acting on us is evoked by many great thinkers, most clearly and articulately by Carl Jung.

TOBIAS: Unquestionably, there are even adverse archetypes and divisions within a collective consciousness as well.

LASZLO: Archetype is the key concept. Our everyday consciousness is but the surface of our experience. There is more to our mind and consciousness than that. And this is why I don't see ourselves being constrained by the current situation.

TOBIAS: So we are not constrained, but that doesn't contradict the existence of the crisis we are discussing.

LASZLO: It does not, but it empowers and potentiates the possibilities for overcoming it.

TOBIAS: You have mentioned the anthropologist Margaret Mead.

LASZLO: Yes, her famous pronouncement is very relevant. A small group can indeed change the world, as Margaret Mead said. And how big that group is depends on the resilience of the system that dominates the world. This seems like a paradox. The more resilient the system, the larger the "small group" has to be if it is to change it, because it has to overcome the system's defenses. For rapidly changing a system, we must have a system that is on the point of collapsing. Then we can create the right kind of fluctuation, inject the right kind of information. All we need is a small group, a small "critical mass."

Of course, that group can change the system not just for the better, but also for the worse . . . as we have seen time and time again in the course of the twentieth century.

TOBIAS: If you look at such worst-case scenarios in terms of a very small mob, as Elias Canetti so brilliantly analyzed in his remarkable 1960 book, *Masse und Macht,* or *Crowds and Power,* in the case of Weimar and then Nazi Germany, you had an unfathomable level of human cruelty occurring that defies all of our mathematical certainties, psychological profiling, ethnographic anomalies, and scientific reasoning. We are left abandoned, scratching our heads, losing faith in everything, shorn of any motivation, hope, or potential therapy. The Nazis proved that time cannot heal wounds—that wounds themselves turn gangrenous and that only the most overwhelming belief in forgiveness allows us to move forward.

LASZLO: The Weimar Republic was ready to collapse when Hitler climbed on a beer barrel in Munich. Czarist Russia was ready to collapse when Lenin showed up in Saint Petersburg with a handful of followers. At these points, change could take place in the system, because the system was not sustainable. Hitler and Lenin, and then Mao and Castro and others did not create the change, they only oriented its outcome.

TOBIAS: And in Japan the crisis was the scarcity of oil that prompted its own spiral into madness, resulting in, one could easily argue, Pearl Harbor and the reactions by Roosevelt and ultimately Truman. I think in microcosm it's interesting what's happening on a place like Easter Island, or Rapa Nui. Today, with a total population of approximately three thousand people and a disastrous legacy (one that is cited more frequently than any other as the ultimate in a collapse of a civilization), the Rapa Nui people are taking all of the best aspects of their cultural norms from hundreds of years of far-flung ancestors who arrived there during the Polynesian out-migrations in the early Western Middle Ages, and they have managed to nurture a virtual Renaissance, ecologically speaking.

Children have been taught principles of ecodynamics, how to avoid the crisis that wiped out their ancestors (as many as fifteen thousand people during the seventeenth century), even down to planting native species on volcanic slopes to essentially collect microclimates of shadow moisture, cultivating precious dew drops and engendering nurseries for endemic species. And they are into it. They are bringing Hawaiian schoolchildren to Rapa Nui to share strategies, noting that Hawaii is the most critical biological hot spot in the United States. It has more endangered species than any state in America, followed by California and Alabama.

And so here we have this critical example of contemporary exchanges that are very down to earth and suggest that we do, in fact, have what it takes to accelerate our own survival through dialogue, through listening to one another, and searching for the best practices in every sector of human behavior and intelligence. Searching across social media, instantaneously all over the world wherever WiFi exists. I suppose I'm trying to suggest that this means, as you say, that we can hope for a reversal. We can tend our garden à la Candide (on the legendary outskirts of some Constantinople), only too aware of all those incipient capacities inherent to the state of human civilization wherein it could easily wipe itself out, or just as easily jump-start a renascence. It's quite a polarity if ever there was one. Perhaps, in fact, a contagious ideal infects a critical mass and we make it through this pandemic Scylla and Charybdis.

LASZLO: We need a transformation from the ground up, triggered by a critical mass. Today's global system is unstable, but we don't know just how unstable it is, so we don't know the size of the critical mass needed to change it. I mentioned yesterday that there are people who will not change, people with a serious vested interest in the current system. There are also people who live at the lowest bounds of survival in deprivation and poverty—they cannot change. But in between these two extremes there is a critical mass that is capable of changing and is open to changing.

TOBIAS: I like that idea of a middle ground. I also think it is a very realistic assessment of the two extremes, as tragic as it sounds. In reality, given that we are very much aware of the billions of our fellow beings who are suffering and the trillions of others, particularly amongst the aqueous biomes, the hundreds of billions of slaughtered beings. I myself focus continually upon new ideas for amelioration of pain. I particularly think about how we can encourage love of nature in others to the extent that they don't simply pay lip service to it, but become the change about which their ideals hopefully whisper in their hearts.

LASZLO: They respond to the ideal of universal love.

TOBIAS: Absolutely.

LASZLO: And this is actually occurring. It is the "cultural mutation" that is already underway. It merits taking very seriously.

———◆———

Afternoon, in the Garden
Overlooking the Valley

TOBIAS: We were speaking of love and cultural mutation, and here we are now in a shed meant for meditation, with the statue of a young sitting Buddha overlooking the Cecina Valley. Meditators and nature lovers are increasing in numbers: seven hundred million people in the US visited the national parks last year, all with the vivacity and excitement, the sheer lyricism of, say, an oboe concerto by Albinoni; something grand and wondrous, not just for the children but for everyone.

LASZLO: And the number of such people is up from previous years I suppose.

TOBIAS: Yes, it is. Also because it is less expensive to visit Yosemite than Chamonix, if you happen to live in Fresno.

LASZLO: Moreover mobility in the world is increasing with the lower costs of travel.

TOBIAS: There is also the phenomenon of Kindle. My mother travels with two hundred books in the palm of her hand. It's a revolution we haven't seen, really, since the time of John Caxton and of Gutenberg. Two hundred books now weighing virtually nothing, with no fear of arthritis in the hands of an elderly but passionate and inspired woman. These may sound like rather mundane things in comparison to the enormity of crises we've been discussing, but not really.

LASZLO: This is all part of the potential for rapid change. That's the potential we are talking about, in what we place our trust.

TOBIAS: If I put on an optimistic demeanor, which I can and would prefer to do, believe me—the upbeat quality, say, of Mozart's *Jupiter* Symphony, which of course he didn't know would be his last—then I can nix the dark side in myself like that [snaps fingers]. Empirically I can bring up so much information in myself that persuades even me that I'm wrong about the darkness. And I love it when I do that. Even in my sobering reflections about the state of the world, especially sober, I crave coming upon an argument that persuades me to cool it, calm down, smell the flowers, and smile. I just am trying to be guarded in my optimism, I suppose. I don't want to make a mistake and suggest that everything is okay. Meteorologists have the best of jobs: when they're right, they are geniuses. When they are wrong, well, everybody knows the weather is fickle.

LASZLO: Complacency is a very real and very serious threat.

TOBIAS: Indeed.

LASZLO: it is just as much of a threat as pessimism and fatalism. They are all dead ends.

TOBIAS: Yes. My worst nightmare is to assume that the party is for me and to think that when I open the door people are going to surprise me. By that I am implying that my biggest fear is suggesting to one I love that everything is fine when I know, in fact, that we are clinging to the wall with our fingertips. So we want to balance but humanity is truly on a tightrope. This was essentially the title of a book coauthored by Paul R. Ehrlich, the Chairman of Conservation Biology at Stanford University and one of the most brilliant scientists of this or any generation, and his equally mindful colleague, Robert E. Ornstein. We are on a tightrope like never before. We have to be very careful.

LASZLO: But there is continuity supporting us. That is a very important factor. If we had to start over again making mistakes and learning from them, every short lifetime would start like a newborn babe. Then the chances of finding the right way forward would be minimal.

TOBIAS: That would be challenging. But learning from past lives is a fascinating prospect.

LASZLO: We can do that, because there is uncanceled memory, total continuity in the subconscious regions of our consciousness. What is evolving on this planet is the same consciousness that is evolving throughout the universe. Here on Earth this consciousness is moving to higher levels. This can overcome the fractioning and the infighting created as its by-product.

TOBIAS: Well that's a positive, as it were; that's a very positive, hopeful window on where we can go. Because there is no question that we have, in many respects, much to be proud of as a species, in spite of the disasters, countless calamities, and disappearances of those twenty-two civilizations the historian Arnold Toynbee dissected. We indeed retain certain meritorious underpinnings in our psyches. I mean you can't go to Chauvet and not imagine that we don't have some kind of blessing bestowed in us.

Or as I mentioned earlier, there is the amazing case of Shanidar 4, where about sixty thousand years ago, in what is today the Zagros Mountains of northern Kurdistan, someone was buried by another human and a tiara of dried flowers placed on the skull. It was an emblem that is potent with our power of conscience, of acknowledging cultural continuity between generations in a region that comprised the greatest accretion of vastly different Paleolithic and Mesolithic cultures and cultural manifestations and populations all cohabiting the same caves and mountain deserts together for many millennia and possibly the place of the first temple, the first church.

The seedling of love in what we term the family or community unit, in fact. A startling icon at the root of our social interactions, certainly at the level of kin altruism, however one argues the empirical evidence which, in the case of Shanidar, first came to light as a result of an archaeological expedition in the mid-1950s by scientists from Columbia University in New York. The fact of the discovery of the tiara was subsequently debated. Some argued it might have been accidental shifts in bird droppings, seismic or drainage fluctuations carrying other species of desiccated flowers, so that the whole conscience theory is mistaken. But even if that turns out to be true, there are other aspects of the caves at Shanidar that unambiguously point to the coexistence of Neanderthal and Cro-Magnon, in other words *Homo* subtaxons, living side by side presumably in peace, and later, of many of the earliest indigenous spiritual traditions manifesting their temples and sacred spaces all adjacent to one another and living in tolerance. Empathy in action, communities of communities.

LASZLO: There are close relations not only among the living, but also with the dead. Ancestor worship is an age-old practice. It may be that worshiping our ancestors is really worshiping ourselves in one of our previous existences.

TOBIAS: That could be . . .

LASZLO: Indigenous people are worshiping their ancestors because they know, instinctively and intuitively, that they are not just their offspring, but because they are them—lived in a former life.

TOBIAS: And conversely, when we look to the future I think we are shepherding that future now; I think we are recognizing it. Scientists and poets alike are peering down the looking glass, in some ways bearing the brunt of emotional and spiritual duty in the guise of hosting responsibility for the future like never before. And this is called stewardship in its most basic sense. Stewardship of the now and of tomorrow—the intergenerational value-laden passing of the torch inherent in most theories and practices of law. US Supreme Court Justice William Douglas's two-decades of personal activism behind the 1964 Wilderness Act said it all when he espoused legal rights for fish, for trees, for watersheds, as well as for people.

And it is, I think, inherent biologically to our species, and certainly to all mammalian species, to pass along mother's milk in whatever form is realistic for a given situation. I can't think of a single mammalian species for whom nurturance is not a fundament. Among the more than five hundred known species of mammals, we see that the exercise of this quality is crucial. It helps define us, not that we actually need definition. But this characteristic, widespread throughout all living beings, is amplified in some ways in mammals and it is truly beautiful. Speaking of a contagious ideal, this is one— laid out for us bare, naked, and unblushing, as Goethe described it speaking in a different context but with words applicable to our remarkable distinguishing characteristics. A major truth. We do have across this planet a framework for moving forward.

LASZLO: We do. This is something that is unfolding, and unfolding throughout the realms of space and time. Today we can speak of the universe as an integral domain, a coherent system in its own right, with all things harmonized within it. In the past, this was but philosophy, not physics but metaphysics. Now we are beginning to see that this is the most rational explanation of our observations. Of the way things are.

TOBIAS: It would be so interesting if Einstein were alive today. What would he think, say, do . . . or Leonardo da Vinci.

LASZLO: Regarding input from Mr. Einstein, did I tell you that I am receiving messages from him through a medium—

TOBIAS: Really! And?

LASZLO: A well-known medium came to see me when I was in Japan to present the Fuji Declaration at the foot of Mount Fuji. This gentleman said that he was advised to come to me by "Professor Einstein" and "Mr. Nostradamus." He handed me long messages from them. According to Professor Einstein, the theories I am developing are on the right tack, but need to be completed—there are ten dimensions and not four.

TOBIAS: How did you respond?

LASZLO: I encouraged him to keep sending me the messages, I am interested. And he does. I get a new message every so often.

TOBIAS: Beautiful. Your discussion in your latest book of the channeling of the deceased Bertrand Russell was fascinating, I must say. It was extraordinarily interesting. It would be foolhardy to be skeptical when, if anything, this is free medicine; accept it or not.

LASZLO: I do believe that there are periodic discontinuities in our bodily existence, but there is a multiphase continuity in the existence of our mind and consciousness. This insight is not new. It appeared already in the Upanishad and in other Vedic scriptures.

TOBIAS: That's right. It appears in the cosmologies of Jain tradition, not to mention most indigenous spirituality and their world visions. A film production of ours once spent nine days in Arnhem Land in northern Australia with a group of Aborigines doing a once-per-half-century recreation of the obtaining of fire, one of their most sacred rituals inherent to all cultural evolution, but was there manifested in a specific norm of acquisition that seemed to speak to all times, all needs, the basis of human evolution. They performed these amazing rites, which members of our team had the good fortune of filming for over a week. And when you consider the implications of fire for our species, you could see the importance to these people, a core value of survival dating back thirty thousand years.

For our predecessor species of *Homo*, the current belief amongst paleontologists is that we acquired fire approximately 1 million years ago at the Wonderwerk Cave in South Africa, and possibly 1.3 million years ago at a series of fire pits east of Lake Baringo in modern day Kenya called Chesowanja. But to see this actual transmission of a mystical experience in Northern Australia was utterly astonishing. These people are us; their journey is our journey. Their discoveries, our discoveries. We were so fortunate to have been allowed to share their beautiful insight into human nature and survival with the entire world through their TV screens. The Australian Aborigine generosity is contagious: They are magnificent exemplars of that cultural paradigm which is largely sustainable and ethically in touch with their environment.

LASZLO: They live the mystical experience. It is their lived, immediate experience.

TOBIAS: That's right.

LASZLO: I am quoting towards the end of my book an important conclusion from *The Golden Bough*, the masterpiece of the famous anthropologist Sir James Frazer. He says that the so-called "primitive" has a deeper, more extended consciousness than modern people. It appears that there is a sense in which we modern people have forgotten wisdom we have had at the dawn of civilization. We could well ask if the advance of civilization is really a progress?

TOBIAS: Of course. Absolutely. I mean, my God, I wouldn't know where to begin in that arena. I have had the good fortune in my life to live with indigenous peoples, many different groups, and to walk with them, and it does surprise me the extent to which modernity is so quick to obfuscate the lessons of our past as well as our present. It's as if this universal consumerism in our mailboxes was shipped overnight. This all expedient consummation of our allegedly common consumerist aspirations has transcended everything our species has otherwise learned since its first defining parturition, defining with a single offspring the anatomy of *Homo sapiens*. You see it in every identical airport in the world, which have all been built by two or three different firms. But the homogenization of the global commons is so pervasive . . .

LASZLO: If you drive to the international airport in Budapest, the last area you find is the shopping plaza. And though that is still in Hungary, there isn't a single Hungarian shop on that plaza. It could be anywhere.

TOBIAS: Diversity is breathtaking; homogeneity and commoditization are frightening. We spoke earlier of cuisine at the table. Your life partner, Carita, is pure genius in the kitchen, as Italy is pure genius in its allocation of time in the early afternoon for a culinary repast which, I am happy to confess, admits to locally-grown organic wine from your own estate, the height of decadent sophistication and telluric blessings. Thinking of all the wonderful different ethnic and culinary traditions, demeanors, and gestures, I don't think we even have the capacity (should we ever be mad enough to want it) to lose that level of diversity. For one example, look at just how painful it was to lose some of the currencies when the Euro in the year 2000 suddenly replaced the French Franc and the Dutch Guilder. It was really I think a tremendous loss to human multiculturalism and art, but there it is, there it was and, so far, there it remains. What we were discussing earlier—I don't think we are going to have the time to lose the fundamental instinctive diversities you intimated. We are either going to get it right or we are not. And, I suspect, the time frame in which we get it right will pretty much determine that we'll retain most of the intercultural diversity. I don't see the new libraries of Alexandria burning anytime soon, although I find it alarming that libraries everywhere on campuses are deaccessioning resources for purposes of digitalization. I think, if anything, the acquisition of instant access to data is thrilling, and I see that as a major positive element.

All this gives me some intimation of hope. We had great debates when Paul Ehrlich and I wrote our book for the University of Chicago

(*Hope on Earth: A Conversation*). We asked ourselves whether or not there should be a question mark following the title. In the end we said, no, let's get rid of the question mark and call it just *Hope on Earth*.

Laszlo: In such questions we must all trust our own judgment.

Tobias: I'm thinking of physics. How does physics, in a general sense, provide insights for humanity? It's a big question, I realize.

Laszlo: It all hinges on discovering the real nature of evolution. Without evolution, there would not be different segments of reality in the world, and without assuming that they came about in the course of evolution, we could not understand how they came to be connected. But if we believe in the reality of evolution, we can explain how physical levels in the world lead to chemical, biological, biosocial, cultural, and psychological levels. Once a new level of evolution emerges, it is not reducible to the lower levels. It is not true that all the complexity in the world can be explained by complex relations among unchanging simple parts.

Tobias: If evolution has consciousness-related and consciousness-explicit phases, at what point is humanity now in terms of physics, and hence of our evolutionary movement as defined by physics?

Laszlo: All that we find in the world now has existed in the past, if not as reality, then as a potential.

Tobias: So where then do we begin the explanation for our humanity, given the fact that in the blink of an eye, fifty million years ago, what is today the state of Montana came to be. What then can best, in your view, explain who we are?

LASZLO: The explanation is in terms of process, of becoming. Evolution is the key factor in our understanding of what the world is, and of who we are.

TOBIAS: And more specifically?

LASZLO: More specifically, this evolutionary process reaches different levels. It reached the level of living organisms after some four billion years of evolution. We have reached now the level of complex living organisms, and the associations and communities formed by these organisms. We call them ecologies, or economies, or cultures and societies, or nations, depending on the aspect under which we view them. In terms of evolutionary unfolding, these are superordinate levels because the elements of these systems are themselves living systems. But living systems are not just assemblies of atoms, and a human society and an ecosystem is not just an agglomeration of individual organisms. Systems on every level evolve their own characteristics. The characteristics at each level are not reducible to the characteristics of systems at the lower levels.

TOBIAS: So do you see disassociation occurring as a normal spin-off of this evolutionary process? I mean, not reducibilities; we know that extinctions are disassociations, and we know that fragmentation impedes any kind of unity. It certainly does on the genetic level. I wouldn't know on the consciousness level. Even in the most pragmatic sense, when we speak, for example, of barriers to dispersion and barriers to distribution in any given area, as much in Italy as in eastern Peru or the Celebes Sea, a single mountain range presents a barrier to distribution of certain species, which can lead to a genetic cul-de-sac. Or it can lead to a mountain pass that might, with a little help from geology, suddenly provide migratory pathways for butter-flies or for ladybugs, as happens so frequently in minute distances on 14,000 foot passes in the Colorado Rockies—to take but one out of millions of examples. Or some ancient Greek or Macedonian phalanx of soldiers marching into battle, easily tracking invasive grass species in the mud on their boots across riversheds . . . the point being, do you see the breakdown, which in a grander sense is the peril we have discussed, at present, and the peril that humanity has felt as a composite part of this evolutionary uplift?

LASZLO: Evolution is not a smooth, uniformly unfolding process. It is strongly nonlinear. And it is not uniformly continuous; not every branch is certain to evolve. Some can disappear. Species extinction is not an exceptional phenomenon, it is even a frequent phenomenon.

TOBIAS: Yes.

LASZLO: Evolution continues despite these disappearances, these discontinuities. It continues, and it explores all possible avenues for its unfolding. The planetary biosphere is not the best of all possible worlds in regard to the continuous evolution of species. Every system here is exposed to the requirement of adaptation to the other systems—and risk of extinction.

TOBIAS: It's interesting you have integrated Voltaire's interpretation of Leibniz in intimating that aspect of the best of all possible worlds, which is virtually a utopian fiction at this point if you drive down any street in downtown Calcutta or most of Detroit. As human beings we have this persistent hope, which we discussed at some level yesterday and the day before. For you personally, what are some of the real precursors for the positive change that you'd like to see? I mean you have an ongoing career that is diverse, to put it mildly, and in depth, obviously, and what are some of the most high-voltage revelatory moments for you that have led you to this point in your thinking?

LASZLO: First of all, I should recognize the role of intuition in my thinking. My thoughts are not uniquely based on lived experience. It's not what I experienced at any given time that gives me the key idea, the master key to what I develop subsequently. Rather the key is the search for the implications of the things that I understand intuitively. This leads me to search for the things that I don't yet understand.

Creating theories is both an intuitive and a rational process. The intuition comes in seeking the understanding of some questions or experiences and knowing whether I am on the right tack. Does it feel right? That is the intuitive side. But the intuition also has to work out. It has to be the simplest consistent explanation of the facts I am concerned with. That is what I am searching for. The fact that I am searching for this is an expression of a drive or impetus that is in me, but not only in me. It is basic to life. It is the search for finding unity, finding coherence. I like to use the term proposed by Stanislav Grof: "*holotropic.*" It is made up of *holos*—meaning wholeness or whole in Greek—and *tropein*, meaning oriented towards something. In this case, it indicates an orientation towards wholeness. Evolution, we can say, is oriented towards wholeness. Among all the alternatives it finds, it selects those that enhance wholeness. This orientation is towards the coherence of diverse elements, and not towards the accumulation of the same or same kind of parts. Evolution is not towards uniformity, but towards the harmony among diverse elements.

TOBIAS: Yes, that is a powerful definition.

LASZLO: Coming together as integrated systems is the purpose that underlies the evolution of living systems.

TOBIAS: Which in turn prescribes almost a given truth with respect to the whole Gaia hypothesis. If you take the entire diversity and multiplicity of biodiversity and geomorphology, hydrology, atmospherics, et cetera, and place it in a single planet, call it Earth, and suggest that she is a living body and there is a wholeness and a unity within the diversity, then you've got, in a very real sense, a perfect expert system, I would think. Isn't that the essence of religion, of spirituality?

LASZLO: That search for wholeness is also exhibited in religion and in spirituality. It is basic to the human mind. It comes to the fore in lived, experienced faith and spirituality, rather than in religious and spiritual dogma. What we experience at the deepest level is the experience of what we are. It is the experience of an entity shaped by the holotropic drive toward integration, oneness, and unity.

TOBIAS: So basically what you're saying, as I understand it, is that the veils of perception belie a "meta" category of being, something bigger than itself. The truth lies somewhere behind that. Thus, even to speak of ecology and all the subsets of those natural history disciplines is, as Planck said, ambitiously off the mark or tangential to the thing-in-itself. We simply cannot get hold of any existential basis in reality. We can only express the facets of it, but never fully get behind it. The pure basis eludes us. We know this paradox from countless insights, be they Sufi, Taoist, Jain, or indigenous North American; from the incredible power of all art forms and the alluring poetics that we have come to deeply sense within the worlds of biosemiotics—interspecies communications. And ultimately we know this hidden reality, in part from our as yet limited grasp of physics and the laws of the cosmos, as well as the increasingly bewildering and magnificent imagery of the universe.

LASZLO: Everything we truly experience of the world results from holotropic seeking. The more we realize this, the more effective is our search and the more articulate our experience.

TOBIAS: Hence that famous Einstein quote about trying to solve problems with the same mindset that created them in the first place.

LASZLO: Yes. We created the problems by forgetting who we are and thinking of ourselves as mechanical aggregates and of the world as a super-mechanical aggregate. This kind of thinking is bound to go wrong because it ignores the interconnections that make the parts into parts of the whole. With reductionist thinking we see the parts and don't see that how they are related together is as real as the parts themselves. What a thing is, as Whitehead said, is decided by its connections to all other things.

TOBIAS: You're really describing a huge gap between a mechanistic viewpoint and a holistic, organic viewpoint.

LASZLO: That gap is the crux of the problem.

TOBIAS: You've been restlessly and methodically (a hard thing to do at the same time) striving to achieve this new paradigm, a new way of seeing that can get us clearly in a multidisciplinary approach to that which is beneath the pressure or the goal or the engine, the fuel that is pushing this whole evolutionary phenomenon that we are experiencing and expressing and groping to express. You've articulated that perpetually, and now do you find yourself at some particular juncture that you would see as different or suddenly shocking within the span of your own lifetime?

LASZLO: New things are emerging also as a result of pursuing a train of thought which itself is not new. But if you pursue it and try to match it to the reality you observe, you find new things, unexpected things, not only things that have already been observed and documented.

TOBIAS: But is there not something absolutely distinct at this point that you feel is emerging?

LASZLO: It is emerging but not just for me, and not just as a result of thinking. I would like to remain with the ideas that have emerged for me, but every time I look at what I have written objectively, I find that I can improve on it, and that there is a better way of expressing what I want to express. I have not reached the point where I could say, "Okay, this is good, now I can let it go." Sometimes I would like to reach that point.

TOBIAS: But wait, you're speaking of yourself. We've spoken of bifurcation, which you've elaborated upon greatly, and we've discussed of course this crisis that I am coming from with respect to our trying somehow to wrap our collective thoughts around the global ecological disasters stemming from our species' unanimous behavioral patterns, be they anomalies in terms of what evolution requires, expects, or can withstand, or not. I'm wondering if at this juncture in your own ontological connection to this worldview that you've addressed—are the underpinnings that you now sense as coming together from the recent revelations in physics and cosmology and contemporary philosophy presenting you with some kind of challenge that you feel is unique to this time? If I asked that same question on the evening of September 1, 1939, the day Hitler ordered the invasion of Poland, you'd obviously have some kind of response. And perhaps also if I asked it on the eve of the year 1000, when everyone within walking distance gathered at St. Peter's Square, thinking it was the end of the world, goaded on by fortune tellers and biblical mind readers hot with the anxiety of the apocalypse, not merely the end of the millennium. And yet, after all of that anticipatory anxiety, along came the sunrise, another day, and it was more or less business as usual. In a time of congested daily revelations in the physical and natural sciences and within your own thinking, have you experienced any of those revelatory sea changes and epiphanies that shed light on some new and urgent necessity? Presenting before you a task or threat or perhaps some intellectual or moral opportunity you didn't see coming?

LASZLO: What is coming together is evidence for something that has always been present, but was previously implicit. This is what is happening when we reach a cusp in our thinking, a bifurcation. This, as I said, is a strongly nonlinear process, a process that is full of stop-and-go, of fall back and leap forward. We are now facing a fall-back and wondering how we can regenerate ourselves and a healthy environment from where we are without having to go down the ladder of evolution and start again from simple organisms.

DAY SIX

Morning, in the Study

TOBIAS: The idea that we are capable of being one thing, and then when we are in the collective being quite another thing, disturbs me. These behaviors are rarely reconcilable. Individuals have amazing capacities—endemic and seemingly infinite capabilities. But put them in a collective and it all goes south. These collective predilections, at least if history be our guide, are incapable of manifesting the best of our individual insights, inspirations, and intuitions. But as individuals, humans ask, augment, investigate, and create—they can be beautiful and unprecedented. I think of my own life partner, Jane Gray Morrison. She is a true individual, a beautiful genius, and I am so blessed to be with her for nearly thirty years now. This is testimony, in my mind, scientific testimony to the fact that human collectives, in this case numbering two, can make it work here on Earth, however slightly sentimental that may sound. But never mind. It is a success story counter to the mass detour that seems by and large to characterize the myriad missteps of our collective forebears and contemporaries, notwithstanding all evidence to the contrary—which is to say, here we are, you and I, together on your porch in Tuscany, having a fine conversation on a fine day in that same history. Talk about endless contradictions. Maybe this is just one of those rarified few contradictions that made it out of the fray, a dialectic that escaped the general course of mass destructions. I don't know. Fallout versus escape. Are these characteristics of the bifurcation?

LASZLO: I don't see that it would be the collective that has gone wrong. The collective expresses what the individual thinks and perceives, but not only that. Sometimes it's the individual who doesn't see things going wrong because there are so many ways of interpreting what he or she sees that it is hard to hit on the right one. And a collective has its own intrinsic dynamic, and that built-in dynamic sometimes goes awry and endangers the very process that has created it.

TOBIAS: Yes, but Ervin, wouldn't you agree that politburos cannot create a Gauguin? I mean you cannot write a book or play a piece by Liszt by committee, unless you somehow expand the meaning of the word "committee."

LASZLO: You are putting the emphasis on individual creativity. If you look at it in purely systemic terms, what we have are ever larger systems forming, ever more groups and clusters on various levels being created, both local and global. Diversification must always accompany mere multiplication. It's just as much a part of the process.

TOBIAS: Is it just as much?

LASZLO: Diversification is an aspect of the evolutionary process. I like to think of it in terms of a process that integrates on top and diversifies on the bottom. Look for example at the European Union. You have an integration on the level of nation-states, but there is at the same time diversification on the bottom level, as more and more civic and ethnic groups claim their separate identity. They sometime break up even previously unified nation-states, such as Yugoslavia and Czechoslovakia, and even a strongly profiled nation such as Spain. These are processes of diversification expressing local identity. Both unification and diversification are occurring at the same time. This is the same process that creates integrated yet individually diverse functional units in our brain.

TOBIAS: The most fundamental population dynamic that we know of is boom and bust. And we measure it, and we are quite clear about its mathematical verities. Typically, 90 percent of individuals die out in any three to four year vertebrate cycle, the boreal lemming being one of the most classic cases in point, and quite contrary, by the way, to the Disney version of their ecological fates. For larger mammals there is a different time frame, whether we are speaking of mountain lions on the North Rim of the Grand Canyon or of caribou in Alaska, or wolverines in southeastern Canada. We've seen it with every major epidemiological situation over the span of a generation. If you speak of population dynamics, we're speaking of a huge fallout. But all of that biological carnage, in fact, goes towards sustaining the population, ultimately multiple populations, whose geographic diversity and gene buoyancies are key to a successful ecosystem.

LASZLO: But there is something remaining, even in a fallout. Because a new boom can take off from there, it doesn't need to start from the same level as the previous boom did. This is precisely why there is evolution, a process that is progressive and irreversible, despite occasional reversals. If this were not so, humanity would still be where it was fifty million years ago.

TOBIAS: If you look at most ecosystems that have achieved the stability of boom and bust dynamics described above, you'll see that whether here in Tuscany or in Alaska or any biome on the planet, they've achieved not only stability, but with that balancing act a distinct signature. By that I mean to intimate the richness of biodiversity at levels of choreography, song, intelligence, acumen, learned experience, intuitive grasps, sensory genius—all of those traits we have until recently tended to reserve for our own species.

Lemmings have achieved that stability, these amazing qualia, despite life cycles that sweep their populations every three to four years. Despite the population stability, multiple individuals die trying to swim across waterways in search of food. Sometimes as much as 90 percent of the total population perishes. Of course, if 100 percent should perish, that's it. End of the story. But typically that 10 percent that survives will swim for about fifteen minutes. They do breaststroke style swimming. But add too many stressors like wind or little waves higher than approximately fifteen inches, and they won't make it.

Human development in Arctic regions, climate change—these comprise additional stressors on top of the primordial, in situ stressors characteristic of any specific biome. The whole system is so highly attuned that, as with, say, a composition by Mozart, you remove one or more notes and the whole beautiful edifice starts to collapse. For the lemmings, those that die provide organic remains for a subsequent feeding frenzy that ensues for the Arctic foxes and owls and falcons. It's a system that has been in place for millions of years, and it is a totally stable system. Like an opera by Mozart. Is it evolving? We don't have a time frame to gauge that, just as climax forests are an illusion. Just as human art forms witness evolving styles and needs.

There is ecological flux at the heart of stability. So if you attribute words like clusters and mirror images or echoes of the Big Bang as recurring patterns in the continuing evolutionary energies that are implicit to the biosphere, implicit to consciousness as you have suggested, then we can probably calm down, knowing that the ecological struggles for existence amongst lemmings in some crucial sense mirrors the implosions and tumult that has characterized the universe going back to the presumed beginning—if there even was a beginning.

The details that differentiate mammalian species or, for that matter, all biological species, are rather insignificant in the face of an entire cosmos. That's certainly one way of writing off the mess we have engendered here on this planet. As human beings we're seeing huge fallout in that respect. We're seeing wars, maternal morbidity and mortality, infant morbidity and mortality, all the fallouts from the major killers including heart disease, cancer, diabetes, depression, and obesity, all of these things collectively wiping us out. And this could indeed, medically speaking, suggest that our species has in fact achieved some measure of ecological stability in spite of human-induced pandemics—most notably the Black Death, which was caused by *Yersinia pestis*, and the 1918 influenza epidemic in which I lost some relatives.

And that's it. We're going to stay in this situation and even the crisis points that we have talked about—I could view those crisis points by playing the devil's advocate and suggesting that they are actually a good thing, and that they will in fact lead us to a point where we will survive precisely because we're self-destructive on the collective level. I'm not self-destructive as an individual (at least not on good days), but if we are suicidal at the species level then it could be a fantastic thing, as horrible as that sounds.

LASZLO: The evolution of the larger so-called supersystem may involve the cyclical appearance and disappearance of some of its subsystems. The subsystems may come and go, but the whole system remains and evolves. When you look at evolution on this planet since the first proto-organisms appeared, you find that there is an increase in both diversity and unity even across the periodic disappearance of many of the species and subspecies. Some populations and some entire species are wiped out, but the integration as well as the diversification of those that survive continues. This is the overall process that characterizes evolution. At the biological level it is expressed as the appearance and disappearance and diversification/unification of species, while on the most encompassing astronomical level it is expressed in the appearance and disappearance of stars and planetary systems in the wider context of the evolution of galaxies.

TOBIAS: Does this give you personal solace? I'm curious.

LASZLO: I'm just saying that this is a confirmation that what is happening on this planet is a natural evolutionary process. It involves the die-out of some parts of the overall system. That die-out is not contrary to the overall trend.

TOBIAS: I agree with that 100 percent. My question is whether our consciousness is capable, with all of these wondrous tools and curious goals—utopian and otherwise—that we've conceived as a species, in any way to shape our own destinies within this cosmos.

LASZLO: I am sure we can shape our own behaviors and that how we shape them depends on how and what we cognize and "re-cognize." Nothing is entirely forgotten; everything can be recovered, at least to some extent. That includes our perception of the whole of which we are a part. That perception is not linear and cumulative, it can even reverse. Most so-called primitive cultures have a more acute and developed sense of the whole than most modern people. They have an instinctive, intuitive sense of being part of the whole—of the whole of society and of nature.

TOBIAS: And yet we romance that notion to death. If we ignore the past, all those lost and largely unaccountable populations and civilizations prior to now, and if we ignore the fact that the vast megafaunal extinctions occurred as a result of the earliest settlers in places like Australia and New Zealand, New Caledonia and also Madagascar, then perhaps we can find a false peace for a day or so. But in truth, we cannot ignore these things. Those early settlers in New Zealand and elsewhere wiped out the largest number of vertebrates in the shortest time frame, short of a colliding asteroid, with human hands. We know these things for a fact. England and Ireland by the twelfth century had largely been denuded. The Mediterranean littoral was denuded even by the time of Plato, who commented upon it.

LASZLO: These are particular examples, but the traditional societies that are still extant, like the pre-Columbian Americans, had an unbroken time span of several thousand years. Their time span was enormously larger than the time span of modern civilization. We can hardly even imagine how any civilization could last as long as the Roman did.

TOBIAS: Here we're sitting near the heart of Villanovan and Etruscan civilizations.

LASZLO: These, too, lasted several thousand years before they disappeared.

TOBIAS: Down to the south the Mycenaeans didn't do so badly. They invented the first indoor toilets (excepting those divined at Terra Amata in modern-day Nice) and painted murals graced with dolphins at Knossos on the island of Crete. But eventually these cultures vanished. Even their languages would have vanished but for the work of some tenacious archaeologists.

LASZLO: What we call civilization is a group of people interacting and reproducing according to certain rules and principles that define what is permissible, desirable, and moral. These groups can perpetuate themselves for large periods of time, if not of course forever.

TOBIAS: I agree with what you're saying in the sense that we have the ability and the interest—a lot of us do—to investigate, perpetuate, and learn from the histories of civilization. We know that many of them have gone extinct, while others have survived. We've asked leading questions about the mechanisms involved with success and failure, with collapse and regeneration. The indigenous studies that encompass so many disciplines have really taken hold since people like the hyperpolyglot William Jones, for example, went to India in the late eighteenth century, spoke some forty-one languages, and translated the Old and New Testaments into Assamese and other local languages of northeastern British-controlled India, in an absolute effort to understand the tribal customs of the subcontinent (he's buried in Calcutta). That, in turn, led to such really great minds as Max Müller and Hermann Jacobi, the earliest translators of the Jain and Buddhist cannons.

While we now have an encyclopedic grasp on these languages—we even have the UN involved in trying to preserve languages from going extinct—because so much forced linguistic and cultural assimilation has occurred, from the Caucasus to Yunnan, only something like six thousand human languages are left on the planet. That is not even accounting for the several hundred uncontacted tribes still thought to persist in places like the Amazon and probably parts of West Papua.

All of this said, we have a sense of the values that have been lost or are being lost as rapidly as we are trying to seize them. But I don't think realistically that the First Nations peoples of Canada or the Native Americans are exercising any kind of influence on the practical level, certainly not on the economic or political levels nationwide—on the poetic and artistic and spiritual levels, for sure. If you combine those areas of passion and compassion and interest by today's young people and researchers and viewers worldwide, of course there is a critical bulwark of interest, but it is not going to necessarily save America from co-inducing a world totally out of kilter, which it already is on the ecological front, as we have discussed. But a nuclear armed Iran and North Korea? The latter country's dictator having threatened to turn America to ash? Such neurological whiplashes would unleash unprecedented consequences in the annals of ecological destruction if there were to be another nuclear war. It will not prevent the United States from—you know, you see where I am going with that argument.

LASZLO: I am not suggesting that traditional cultures themselves are a decisive influence. I am suggesting that their example could exercise a decisive influence on people's thinking and values. The truly decisive influence would be the one that is so critical today—namely the re-connection of human society to nature. Traditional cultures had a much closer connection than modern societies.

TOBIAS: No question about it.

LASZLO: And the recovery of that connection on the level of conscious cognition would be a crucial major factor in determining our life and our future.

TOBIAS: I agree.

LASZLO: This rediscovery has been taking place in many parts of the world, but it has been overshadowed by a relentless competition for power and wealth. Many of the cultures that "cognized" and "re-cognized" their close connection to nature became marginalized. But that isn't proof that they were wrong.

TOBIAS: No, of course not. In fact, it is an evidence-based argument for the inevitability of power corrupting and whole civilizations dying out in the struggle to obtain sovereignty. That would explain better than perhaps anything the fate of the Aztecs. It's not that the Aztecs themselves were indigenous peoples romantically connected to the earth—in fact they were anything but. They were no less brutal than the Spaniards. There is little debate that they eviscerated their young by the hundreds of thousands and possibly millions, hard as it is to demographically and morally wrap one's mind around the Aztec realities back in their Middle Ages. They had beautiful, animistic traditions, no doubt about it, and this was reflected in their art and sculptures, just like the terracotta works we see throughout the Levantine cultures, not least those in Cyprus.

But in the greater optics, as you put it quite vividly, all of these conflicts come down to pretty much the same characteristics, with respect to power grabs and ego. Even within those cultures that are so revered today for being closer to nature, they come home at night and wake up to the same problems in the morning—they are human beings. So we are really looking at human natures (plural), which was essentially Indian Prime Minister Nehru's great answer when first asked whether he thought India had a population problem at 300 million. He said, "We don't have a population problem; we have

300 million population problems," or words to that effect. Quite an insightful truism. Every individual counts, and not just amongst our own species in my opinion. Every individual has a role to play to suggest some crucial element in this emerging paradigm of unity, and I have faith in that, actually.

I think we have an opportunity before us that is unique, from my perspective. We have a unique opportunity, I sincerely believe that, and it is the opportunity that has come at the moment we are now unerringly and unambiguously aware of—the moment of our plight. And we want to live, we want our loved ones to live. I see it very practically. I want every living being to continue to be.

LASZLO: The basis of our civilization, the basis of our culture and our worldview, has become moribund; it has to change.

TOBIAS: Yes.

LASZLO: If you view what is happening today in the local and short-term perspective, then you might come to the conclusion that many of the now extant civilizations will go down the drain. But when your view expands to a larger perspective, you will see that what is happening today is part of a cycle, of a larger process in which individual elements, entire civilizations go down and others appear, but the overall process continues. The beat goes on.

TOBIAS: I hope.

LASZLO: What we need for this beat to unfold is, among other things, the re-cognition of what it is that carries it. The basic carrying rhythm is the holotropic movement of evolution towards diversification and integration—towards unity within diversity. The carrying rhythm is the creation of systems within and with systems, producing integrated and yet diversified supersystems. These are the systems that survive and evolve, at least until the time when this planet's biosphere itself disappears. But that, too, is part of the cosmic process, because complex systems cannot persist indefinitely in this universe.

TOBIAS: I was going to say if all was writ from the moment of the Big Bang, or that elusive moment prior to the Big Bang, then the replication of what's happening here theoretically would have happened everywhere that biospheres emerged. And that's also a potential.

LASZLO: The potential is universal, because evolution is constantly operating at the expense of die-outs and fallbacks.

TOBIAS: I didn't mean just that. Truly if, as you speculated, there are other planets with other hydrogen-based life-forms—we are hydrogen-based after all—life-forms as we think of them in terms of ourselves, looking outward, then what's to say that all of the different earths that might be out there, that Carl Sagan first suggested to a global public, aren't all undergoing the same bifurcation points, the same crisis that we are at right now? This is pure speculation. It's fantasy.

LASZLO: The processes of life are consistent in space, but may not be consistent in time. Evolution may have taken off at different times in different places. And it could proceed at different rates, depending on local conditions. A nearly infinite variability is built into the unfolding of the processes of evolution. So it is not very likely that the crises of other civilizations would be occurring at the same time as ours.

TOBIAS: We really know less and less about more and more—so much that is definitive, against that which we can't begin to grasp, let alone experience. And this is another troubling aspect of all this bifurcation/crisis momentum. I find that much of the science today is theoretical and is losing touch with the experiential. That worries me for a number of reasons. Just as the extinction of languages, extinction of experience troubles me. It leads almost overnight to ecological illiteracy, to mean-spiritedness, to lack of parity—not just diminishing gender parity but to oppression in general—and to hatred and all of the negative traits that we see in ourselves. I worry that theoretical knowledge in all of the sciences is, in fact, quite possibly losing touch with a reality that was, after all, perceived rather clearly long before our time.

Science today is not only becoming more and more theoretical but more and more specialized to the point where the holistic points of view are almost marginalized because individual researchers are removed from them. This is true except in cases where you have several hundred, even several thousand scientists collaborating on the same problem, as happens at CERN or the UN talks on Climate Change. I've seen one paper with thirteen hundred or so authors and a relatively recent Nobel Prize in physics that was shared by about three hundred scientists. But they don't have the language of Hegel and Richard Feynman at the same time. They don't have the different languages at their command simultaneously in order to think about these things. Now one could argue that they need to achieve their focal points—as a great opera singer must find her or his true voice—if purveyors of that data are to achieve their own respective focal points. But what worries me is that—and you know this very well—they are struggling for grants, for their own statistical accruals of citations of their peer-reviewed essays in other journals that show up time and again regardless of how significant any specific advance may have been in terms of their specific fields. There is a lot of repetition that is simply predicated upon who is getting the funding.

These are very mundane considerations in the larger view of what we have been discussing, but again they are fallouts from this recession, if you will, in which theoretical science is moving away from experiential data. This may be anecdotal but it is experiential, it is cognized by the researchers who reflect on it personally. Without that personage, that human touch, I deeply fear an isolationism that will translate into further marginalization of our humanity and a deepening crisis in the arts, in our ability to communicate with each other, to live together, to be viable as a species within the far-reaching constraints and dictums, no doubt, of evolution.

LASZLO: What is happening today is the beginning of a paradigm-shift. The world around us is shifting, and our thinking about the world, our perception of what is happening in the world, is likewise shifting, even if it is not shifting in tandem with the world. We need a reliable way of seeing what is happening in the world, a new paradigm anchored in the findings of cutting-edge research in science. The choice of our paradigm will decide whether we see what is happening as a movement towards a positive evolutionary outcome, or as a hopeless struggle against higher odds. We have the facts and the observations before us. How we see them is a question of how we interpret them.

TOBIAS: That is interesting. I have spent my life observing and thinking about my observations. So to be confronted with the notion that those observations, while they may have elements in science, may also be totally wrong, is fascinating and enthralling to me. If I could improbably "prove" Alfred Wallace wrong, I'd probably be inclined to race to publish my findings. Paul Ehrlich said something similar with respect to Darwin. I suppose that passion to forever move forward is the nature of science.

LASZLO: You can't actually prove any theory in science.

TOBIAS: Right.

LASZLO: As the philosopher of science Karl Popper said, scientific theories are not provable, but they are improvable. A nice play on words, and it's true. You can't prove absolutely any statement or observation. There is no such thing as a crucial experiment that decides whether what you propose is true or false. For any finite set of observations there are an indefinite if not infinite number of explanations. Some are more consistent than others, and scientists tend to adopt those. Some are also simpler and more insightful, or what mathematicians call "elegant," and scientists take those for true. Ultimately, it's not the observations themselves that dictate what explanation we choose for them, but our values and preferences. It's the optimally consistent and at the same time optimally simple explanation that scientists look for, and—if they are consistent and unbiased—also adopt.

TOBIAS: I'm taken by what you just said with respect to the variability of interpretations, because the grandest experiments produce seemingly grand results—the special theory of relativity or the second law of thermodynamics, for example. If they were shown to somehow be flawed arguments, would that be because the interpreters were flawed or because the data was flawed?

LASZLO: The second law of thermodynamics is an ideal and not a one-to-one mapping of something that takes place in reality. In fact, it does not apply to the real world because it only applies to closed systems, and there are no systems that would be fully and effectively closed in nature. If we want a true mapping of the real world, we have to use concepts applicable to open systems. When we take the systems that truly exist in the world—which are open system, constantly exchanging flow energies and information with the rest of the world—and close down their connections, they would become closed systems, but that is an abstraction. Processes in the real world are not characterized by a drive or tendency toward maximum entropy because they are constantly and effectively connected with the rest of the world. In consequence, the evolutionary process is not uniquely an entropy-generating process, but a process where entropic and negentropic processes interact.

TOBIAS: One great case would be Biosphere 2 in Arizona.

LASZLO: All living systems are essentially open systems; we cannot close their connections to the world around them without killing them.

TOBIAS: I think we will never master the impediment to recreating wilderness. It is our hubris, dating from day one of this evolutionary experiment, that has resulted in *Homo sapiens*. Terraforming Mars, for example, which has been on the minds of so many. As to the earliest assumptions that you could take, say, ninety representative species from the entire Earth and hope that scientists might manufacture nature as functionally as the planet Earth herself, well, good luck. It certainly did not work out in that planned greenhouse inhabited by a half dozen or so scientists for two years, shut in a little-hole-in-the-wall in Arizona. It was a most fitting emblem for what could happen on other planets. It was ludicrous, and it proved a lot of things to a lot people. Scientific communities certainly benefited from the mistakes made and obtained a load of data, so in that sense, the arrogance was beneficial. But it was, in the larger sense, in my opinion, arrogance nonetheless. It was an experiment that reminded me of Descartes thinking of animals as machines, or the thirteenth century Holy Roman Emperor Frederick II experimenting on children to see if they could be raised without any other human interaction or language acquisition—allegedly in the hopes of discovering the true language of God. Apparently the children exploited in those experiments died quickly.

Such anecdotal thoughts have informed my perceptions in many ways; they live in my heart and mind reminding me every time I'm out observing the world—and I don't even have to be out to be out. I can be in and still be out, as John Muir put it, every time I go out I'm actually going in. Reading the manuscript of nature simply requires thinking, but it also cries out for the abnegation of that arrogance of power which is so easily corrupted when put to the test. And there is no more brutal or potentially revealing test than an experiment.

LASZLO: That is just what I was talking about in the beginning of our conversation, looking for wisdom on the inside rather than manipulating the outside and then looking for confirmation.

TOBIAS: But I would hate to see a world that has achieved a sense of inner coherence at the price of incoherence at the outside.

LASZLO: I don't think we can achieve internal coherence without external coherence—this is a world of interconnection. Every time we try to disconnect from the rest of the world, we run into disaster.

TOBIAS: Best case argument right there for bringing back that concept I described, the Jain/Gujarati ethical decree that basically declares that all life-forms are interdependent and need one another.

Mutual interdependence was proclaimed by Mahavira in the sixth century before Christ.

LASZLO: That is an amazing insight, and not an accidental one.

TOBIAS: We still have that wisdom. It's not a vapid mantra.

LASZLO: It is in us. And it is precisely what we need to recover.

TOBIAS: A Jain temple in California found itself surrounded by a parking lot and fast-food joint and asked the proprietors of the hamburger franchise to at least serve veggie burgers. Apparently it has been something of a struggle but has, as I see it, only empowered the ancient ethics inherent to Jain tradition that much more. Nothing comes easily. You have to stand up and be prepared for a lifetime of selfless service in the spirit of your convictions. It's that clear. There is conflict everywhere, but we do have what it takes to move forward, I believe. Nothing is handed to us on a pewter platter. Alexander the Great only hired soldiers who passed a very odd little test. Applicants were taken to a river and asked to drink from that one and only river. Those who cupped their hands to get the water, he would not hire. Only those who threw their faces into the water were deemed acceptable for the army.

It's a strange and possibly apocryphal story, but speaks to the heart of our situation: Getting into the wild. We have to work for it. In my case, it is one of the reasons I persistently go out—literally. I do so because of the sheer joy of it, the personal pleasure, the gratification and wonder that comes from simply being with birds and insects and wild fellow friends out here in the wild west of the world. Trying, for instance, to help a tarantula migration that may be caught in the crossfire of a roadway. I get out there and stop traffic to save those fabulous spiders in October in Central California. Truck drivers get out of their high-flying cabs and walk onto the road to see what in the hell I'm doing, and when they see the tarantulas marching in single file, they are enthralled and grateful to me for saving them.

Or rescuing Senegalese parrots from fetid cages in an illegal wild animal market place in Mali or Yemen, or saving chickens and roosters in Kenya. Or pigs in China. That gives me a reason to be myself, to discover who I truly am or aspire to be. And the more I see, the more I hear, the more I feel, the more I feel human. The more I am capable of manifesting my humanity, of helping and loving my wife and life partner, Jane. My mother, Betty. My brother, Marc. Being everyday— Jane and I—with a most magnificent and mysterious Macaw whose every breath and bright gaze is the genius, the promise, and the fragility of wilderness here on Earth.

DAY SIX

Afternoon, in the Study

Laszlo: You and I, as all thinking and serious people, are trying to realize our potentials, trying to open up and move beyond our cocoons—both our mental and our physical cocoons. This opening up has been a basic feature of my aspirations all through my life. To understand the world, and my being in the world, is what I have been searching for. I see that you share this aspiration, and so I want to share with you the simplest, most basic insights I have come across all these years. I call this set of insights my "credo." Let me read for you what I wrote on this not long ago, as my closing contribution to our conversation.

TOBIAS: Thank you, Ervin. I would appreciate that.

LASZLO: I began by stating the principles that I have tried to follow. First, as Einstein said, to seek the simplest possible scheme of thought that can tie together what we observe and experience. Second, to make this scheme as simple as possible—but not, of course, simpler. And third, to heed a warning from Plato, who said that all statements about the real world are but likely stories. My credo is the scheme, the set of insights, that makes up the likeliest story.

This scheme, my credo, has three parts. Part one is about the world, part two is about evolution in the world, and part three is about purpose in the world. Let me read you the first set of insights, about the fundamental nature of the world. These I believe are the elements of the likeliest story we can currently tell about what the world really is.

First, the set of insights about what we discover is the fundamental nature of the world:

> The cosmos, the largest and conceivably infinite reality of the world, is an infinite intelligence.

> The infinite intelligence in-forms the finite universe it has created.

> The things and beings we observe or infer from our observations are clusters of vibration in the finite universe, in-formed by the infinite intelligence.

> At different frequencies and wave lengths, clusters of in-formed vibration, the ultimate reality of things in the universe, appear as structures of matter; as aspects and elements of consciousness; and as transcendental experiences and information.

Second, the insights about the fundamental nature of evolution in the world:

> Clusters of vibration that appear as structures of matter and as aspects and elements of consciousness are in-formed clusters of vibration, and they coevolve in the finite universe. Vibrations that appear as structures of matter evolve toward coherence and ultimately to super-coherence, and those that appear as consciousness evolve toward the apprehension and integration of transcendental experiences and information.

> Vibrations that appear as structures of matter evolve intermittently: they are periodically replaced. Vibrations that appears as consciousness evolve continuously, through iterating phases in space and time of the universe, and phases beyond space and time.

Third, the key insight about the nature of purpose in the world:

> The purpose of evolution in the finite universe is the union of the consciousness that evolves in space and time with the beyond-spacetime infinite intelligence that is the cosmos itself.

These statements encapsulate the insights I have reached in my thinking about body, mind, and the universe. I wanted to share them with you in recognition of your initiative in holding these six days of memorable conversations here, in the heart of Tuscany.

TOBIAS: I want to thank you, Ervin, for these insights, and for our entire conversation. And your life partner, Carita, for being so helpful and such a gracious hostess and creating such a marvelous ambiance for these wonder-filled days.

ABOUT THE AUTHORS

ERVIN LASZLO spent his childhood in Budapest. He was a celebrated child prodigy, making public appearances from the age of nine. Upon receiving a Grand Prize at the international music competition in Geneva, he was allowed to cross the Iron Curtain and begin an international concert career, first in Europe and then in the United States. At the request of Senator Claude Pepper of Florida, he was awarded US citizenship before his twenty-first birthday by an Act of Congress.

Laszlo received the Sorbonne's highest degree, the Doctorat ès Lettres et Sciences Humaines in 1970. Shifting to the life of a scientist and humanist, he lectured at various universities in the United States including Yale, Princeton, Northwestern University, the University of Houston, and the State University of New York. The author, coauthor, or editor of ninety-one books that have appeared in a total of twenty-four languages, Ervin Laszlo has also written several hundred papers and articles in scientific journals and popular magazines.

He is a member of numerous scientific bodies, including the International Academy of Science, the World Academy of Arts and Science, the

International Academy of Philosophy of Science, and the International Medici Academy. Laszlo received the Goi Peace Award in 2001, the Assisi Mandir of Peace Prize in 2006, the Polyhistor Prize of Hungary in 2015, The Luxembourg Peace Prize of 2017, and was nominated for the Nobel Peace Prize in 2004 and 2005. He was elected member of the Hungarian Academy of Science in 2010.

Laszlo is founder and president of the global think tank The Club of Budapest and founder and codirector of the Laszlo New Paradigm Research Center in Italy.

Ervin Laszlo's recent books include: *Value Theory in Philosophy and Social Science* with J. Wilbur (new edition), 2014; *The Self-Actualizing Cosmos: The Akasha Revolution in Science and Human Consciousness* (2014); *The Immortal Mind: Science and the Continuity of Consciousness Beyond the Brain* (with Anthony Peake), 2014; *What Is Consciousness? Three Sages Lift the Veil* (with Larry Dossey and Jean Houston), 2015; *What Is Reality? The New Map of Cosmos and Consciousness* (with Alexander Laszlo and contributors), 2016; *Beyond Fear And Rage: New Light from the Frontiers of Science and Spirituality (ebook)*, 2017; *The Intelligence of the Cosmos: Why Are We Here?: New Answers from the Frontiers of Science* (with an introduction by Jane Goodall and afterword by James O'Dea, and contributions by Emanuel Kuntzelman, Kingsley Dennis, Maria Sagi, et al.), 2017.

DR. MICHAEL CHARLES TOBIAS earned his PhD in the History of Consciousness at the University of California, Santa Cruz, specializing in global ecological ethics and the interdisciplinary humanities. His wide-ranging work embraces the global ecological sciences, art, comparative literature, the history of ideas, philosophy, and natural history in the context of a multitude of possible future scientific, geopolitical, economic, and social scientific scenarios.

He is the author of over fifty books, both fiction and nonfiction, and has written, directed, and/or produced over one hundred films in addition to conducting ecological field research in over ninety countries. His works have been read, translated, and broadcast throughout the world and he has been the recipient of many awards, including the International Courage of Conscience Award.

Tobias has been on the faculties of several colleges and universities, including Dartmouth, the University of California, Santa Barbara as Distinguished Visiting Professor of Environmental Studies and the Regents Lecturer, and the University of New Mexico, Albuquerque as the Visiting Garrey Carruthers Endowed Chair of Honors. An Honorary Member of the Club of Budapest, Tobias is a full member of the Russian International Global Research Academy, as well as the Russian Public Academy of Sciences.

For eighteen years Tobias has been President of the Dancing Star Foundation (www.dancingstarnews.com; www.dancingstarfoundation.org) whose mission is focused on animal liberation, global conservation biology, and environmental education.

Tobias' recent books include *Why Life Matters: Fifty Ecosystems of the Heart and Mind* (with Jane Gray Morrison, Springer), *Hope on Earth: A Conversation* (with Paul Ehrlich, University of Chicago Press), *Anthrozoology: Embracing Co-Existence in the Anthropocene* (with Jane Gray Morrison, Springer), and the novel *Codex Orféo* (Springer).

ABOUT ECOLOGY PRIME™ PUBLICATIONS

Ecology Prime™ is a pioneering global collaborative platform engaging students, teachers, consumers, and businesses worldwide in environmental studies and the day's ecological dynamics. Linking all cultures regardless of language, it provides an easy-to-access-and-use multimedia, communications and social connectivity system that doubles as a source for publishing, eco-exploration, and personal wellness via a network of affiliated destinations around the world.

Ecology Prime™ Publishing was created in collaboration with SelectBooks and Waterside Productions.

http://ecologyprime.com